T0202353

Dead Zones

Dead Zones

*The Loss of Oxygen from Rivers,
Lakes, Seas, and the Ocean*

DAVID L. KIRCHMAN

OXFORD
UNIVERSITY PRESS

OXFORD
UNIVERSITY PRESS

Oxford University Press is a department of the University of Oxford. It furthers
the University's objective of excellence in research, scholarship, and education
by publishing worldwide. Oxford is a registered trade mark of Oxford University
Press in the UK and certain other countries.

Published in the United States of America by Oxford University Press
198 Madison Avenue, New York, NY 10016, United States of America.

© Oxford University Press 2021

CIP data is on file at the Library of Congress
ISBN 978-0-19-752037-6

DOI: 10.1093/oso/9780197520376.001.0001

1 3 5 7 9 8 6 4 2

Printed by Integrated Books International, United States of America

Contents

Prologue

I didn't think much about oxygen during the first trips I did on the Delaware River and Bay in the late 1980s. Honestly, I didn't think much at all, except about being seasick. I do okay on some ships, not so on others. I got sick on my first cruise on the *R/V Knorr* in the New York Bight, but was fine a few years later on the *R/V Atlantis*. I almost died on several trips to the subarctic Pacific on the old *R/V Thomas G. Thompson*, but enjoyed cruising from Hawai'i to Tahiti on the new *R//V Thomas G. Thompson* (even though I saw little of the two islands and a lot of ocean in between). What about the *R/V Cape Henlopen*, used for most of my work on the Delaware River and Bay? She was a difficult vessel to love. I always turned green when we sailed out of the harbor at Lewes and into the Delaware Bay where it meets the North Atlantic Ocean. There, the waves and swells aren't big, but neither was the *Cape* with her flat bottom and shallow draft. As if the motion wasn't nauseating enough, there were the ship's diesel fumes, galley grease, and backed-up heads. As you steam north toward Philadelphia, the waves lay down where the Bay narrows and the shore comes back in sight. Soon the salt of the estuary is gone, and the brown of the river begins. As I worried about the trip back into Lewes and my mortality, I didn't think much about those changes, nor do I remember being surprised at the dissolved oxygen levels. Looking at the data now, I can see that oxygen was low in the river over thirty years ago. At that time, as I would later learn when back on firm ground and breathing fresh air, the Delaware was recovering from being a dead zone.

The Delaware wasn't the first dead zone. The term was first used in a front-page article in the *Houma Daily Courier* on July 22, 1985, to describe low oxygen waters in the Gulf of Mexico.[1] Located 85 miles south of Baton Rouge, Louisiana, Houma is the biggest town in the Terrebonne Parish: home to oilmen, alligator hunters, shrimpers, and fishermen. The *Daily Courier* article mentions that the fourth annual Summer Shrimp/Groundfish Trawl Survey, conducted by the US National Marine Fisheries Service the previous month, had found a 2000 square mile region of bottom waters with low dissolved oxygen.[2] At the time, it was the largest low-oxygen area measured in the Gulf and the reason the reporter, reaching for a phrase befitting the specter of massive fish kills and empty shrimp nets, called it a dead zone. Later, that estimate was revised to 3000 square miles by scientists working out of the Louisiana Universities Marine Consortium Research Center in Cocodrie, a settlement located 30 miles south of Houma. Cocodrie is more bayou and salt marsh than solid land, where shrimp boats outnumber cars. In 2017, scientists found that the dead zone was nearly 9000 square miles or 23,000 square kilometers, the most extensive to date,[3] the size of New Jersey plus most of Rhode Island. Long before 2017, the term *dead zone* had been applied to the many other water bodies around the world where oxygen was disappearing.

"Dead zone" is too melodramatic and inaccurate for many scientists who are usually pretty conservative and picky when it comes to writing up their results. We scientists use "hypoxia," a medical term, to talk about low-oxygen levels. There are legitimate reasons to use that word and another term, "anoxia," when oxygen drops to unmeasurable levels. First, "dead zone" gives the impression that the entire water column, from the surface to the bottom, is without oxygen. In fact, in many dead zones, including the one in the northern Gulf of Mexico, dissolved oxygen is low only in the bottom layer of water. An even bigger problem with the term is that a dead zone isn't really dead, at least not totally so. Some organisms do just fine, even flourish, in low-oxygen waters. The burrowing shrimp survives for a while with just a breath of the gas,[4] and other organisms, admittedly only microscopic ones, do best without any oxygen. The Gulf dead zone has more life than many regions of the ocean.

Although "dead zone" may be over the top, it does accurately convey the impact of low oxygen on biota. What life there is in a dead zone is far from the diverse flora and fauna of a well-oxygenated ecosystem. Without oxygen, animals die, only some more slowly than others. And the gas doesn't have to reach dead-zone levels before effects become evident. For every burrowing shrimp, thousands of other animals are like Atlantic rock crab, which are stressed out when oxygen drops only slightly below fully oxygenated seawater. A small decrease in dissolved oxygen can cause problems because even under the best circumstances, water can't hold much oxygen, much less than the levels seen in air. At sea level and 20 degrees Celsius, air has nearly forty times more oxygen in a liter than does water. That's the reason why whales and dolphins with their big oxygen-hungry brains could not exist on just the oxygen dissolved in the oceans. In Herman Melville's *Moby Dick*, published only a few decades after oxygen's discovery, Ishmael says that a whale is a "mammiferous animal," and Melville knew whales use lungs like us to breathe air.[5] Ishmael apparently knew more biology than did the writers and producers of the 2018 film *Aquaman* in which Nicole Kidman and Willem Dafoe are able to speak and move in an ocean without obvious underwater breathing devices. Gills wouldn't have been able to extract enough oxygen from seawater to support the orations and battles waged in this made-in-Hollywood ocean.

How dead is a dead zone? The level of dissolved oxygen used to define hypoxia and for calling a water body a dead zone varies a bit. The most common cutoff is 2 milligrams of oxygen per liter, although for this book, the exact value doesn't matter much. One way to appreciate those milligrams per liter is to express oxygen relative to "saturation," the maximum amount of a gas that water can take in from the atmosphere and hold as dissolved. Water exposed to air for sufficient time is 100 percent saturated. Hypoxic seawater at 20 degrees Celsius (68 degrees Fahrenheit) with 2 milligrams of oxygen is only about 27 percent saturated. That's less oxygen than in the "death zone" on Mount Everest, where the gas is 40 percent of the amount at sea level, and climbers die while waiting for their turn to reach the summit. Even water at the surface can be undersaturated or less than 100 percent saturation, if oxygen is used by animals and microbes faster than it diffuses in from the atmosphere or is made by algae during photosynthesis. Diffusion of dissolved oxygen from one water mass to another is also very slow, explaining why bottom waters of the Gulf can be hypoxic even when dissolved oxygen is plentiful at the surface.

I didn't know anything about burrowing shrimp before working on this book, but of course, I knew about the importance of oxygen. I had done my share of Winkler titrations and mass spectrometry injections to measure dissolved oxygen levels in many oceans and the Delaware estuary. I did those measurements to study the metabolic activity of microbes and to explore rates of organic matter degradation. Ecologists also use the production of dissolved oxygen to estimate primary production—the growth of plants and algae. Information about these rates are crucial for understanding the carbon cycle. Even just the level of dissolved oxygen is informative. Water supersaturated with oxygen (greater than 100 percent saturation) indicates high algal growth and photosynthesis, which occur during harmful algal blooms. In the other direction, water below 100 percent saturation, an oxygen deficit, can indicate problems. The oxygen deficit was one of two measures used recently to evaluate the effectiveness of the 1972 Clean Water Act, one of the most important pieces of legislation safeguarding the environment in the United States.[6] Even the Act's other measure, a "fishable" standard, was based on the oxygen deficit and a measure of oxygen consumption. If you could measure only one thing to assess the environmental status of a habitat, it would be oxygen. It's an important sign of ecological health. Of course, it's also an important sign of human health. During the month before her passing, my mother talked about her blood oxygen level as a way to quantify how she felt, as if her shortness of breath and weak heartbeat weren't enough. People who have come down with Covid-19 know all too well the importance of oxygen. The American Lung Association's saying, "if you can't breathe nothing else matters," is about us, but it's also applicable to aquatic animals.

So, when oxygen declines, it's a sign of trouble; klaxons and sirens should sound when the decline reaches dead-zone levels. I want this book to sound the alarm about the increasing number of new dead zones and expansion of old ones around the world. But it was the scientist in me that got me started on this book. I wanted to understand the low oxygen levels in the Delaware I saw on those trips in the late 1980s fighting off seasickness. How did the Delaware River get to be a dead zone and how did it recover? What's all the fuss about the Gulf of Mexico? What about other parts of the world?

This book is what I've learned.

1

The Great Stinks

The list of ways to measure dissolved oxygen has to start with the Winkler method developed by Ludwig Wilhelm Winkler in 1888. After being tweaked by J. H. Carpenter in 1965, it is now always called the modified Winkler method. Still in use today, the method is the gold standard for dissolved oxygen measurements. Other, more recently developed methods use Clark electrodes, optodes, or membrane inlet mass spectrophotometers. They vary in electronic and electrochemical sophistication, and all have been computerized, of course, even the modified Winker method. Then there is the human nose. An Israeli colleague once told me that the nose is an excellent microbiologist because it can detect many gases produced by microbes, especially those released when oxygen is absent. Even a stuffed-up nose could have sensed what was happening in the River Thames in the 19th century. Although we don't have data about dissolved oxygen from that time, we know it was a dead zone because it stank.

The stench was so bad that it made the newspapers, even though much of 19th-century London was malodorous from dung-littered streets, sulfurous coal gases spewed out by houses and factories, and thousands of people who could only occasionally bathe. In July 1855, the great scientist Michael Faraday took time out from making fundamental discoveries in electrochemistry to write a letter to *The Times* about his trip by steam boat on the Thames (Fig. 1.1): "The appearance and smell of the water forced themselves on my attention. The whole of the river was opaque, pale brown fluid. . . . The smell was very bad . . . the whole river was for the time a real sewer."[1] A few years after Faraday's letter came what the popular press dubbed the Great Stink. The summer of 1858 was the hottest one on record and so dry that the Thames was much lower than normal, leaving it more incapable than ever of absorbing London's sewage. Another letter to *The Times* summed up the situation well:

Sir: The stinks of our once noble and respected father Thames are now too nauseous for endurance. I rowed on Saturday from Westminster to the Crabtree in Putney Reach and there was no intermittence—no

FARADAY GIVING HIS CARD TO FATHER THAMES;
And we hope the Dirty Fellow will consult the learned Professor.

Figure 1.1 Michael Faraday presenting his business card to Father Thames. This cartoon, originally published on July 21, 1855, in *Punch*, refers to Faraday's letter to *The Times* describing the river's stench.

cessation—one fatal, horrid, open, deadly cesspool. What is to be done? Signed, Oarsman.[2]

The river's effluvium reached higher persons and places than the Oarsman in his wherry. In the newly renovated House of Parliament on the Thames, Benjamin Disraeli fled from the House of Commons with a handkerchief over his nose, bent over, running away from the "pestilential odor." He called the river "a Stygian pool, reeking with ineffable and intolerable horrors."[3] Curtains of Parliament on the riverside were soaked in lime chloride to combat the odors. Queen Victoria and Prince Albert tried to take a pleasure cruise on the Thames, but within minutes were turned back by the stench.[4] As London grew to be the capital of an empire and its population exploded into the millions, the river had become an open sewer.

Sewage's bouquet is due to a mixture of about thirty chemicals, including ethyl mercaptan (which smells like decaying cabbage), scatole (feces), and cadaverine (rotting meat).[5] When dissolved oxygen is abundant, these chemicals are easily degraded to carbon dioxide and other odorless chemicals by bacteria with oxygen's help. One gaseous chemical giving sewage its unique redolence of rotten eggs is hydrogen sulfide. The presence of hydrogen sulfide is a dead giveaway that dissolved oxygen is gone. It is produced only by organic material-degrading bacteria when they have to switch to sulfate after oxygen is depleted to zero. Much of the 19th-century River Thames was anoxic (no oxygen), not just hypoxic (little oxygen), evident from the rotten egg gas bubbling up from the river and mud.

Hydrogen sulfide is produced in nature without sewage's help. Instead of the sewage organics, natural organic material is degraded by sulfate-using bacteria in the absence of oxygen, emitting hydrogen sulfide as a byproduct. I admit to liking a small dose of it when I cycle past my local salt marsh early in the morning. You have to catch it early, before photosynthesis by marsh plants cranks up and oxygen is released, shutting down hydrogen sulfide production. Whether you like a small dose or not, we are quite sensitive to hydrogen sulfide for good reason. The human nose can sniff out as few as ten molecules out of a billion molecules of air. When hydrogen sulfide levels increase by about hundred times, we get headaches and sore throats; when levels reach ten thousand times higher, we die.[6] In his entertaining book, *London Under*, Peter Ackroyd reports that two men entering a London sewer were instantly killed, probably by hydrogen sulfide.[7]

Another sure indication that the Thames was a dead zone is in the fishing reports from the time. As people retched from the river's odors, once abundant fish went missing.

It is hard to imagine the current denizens of Mayfair and Kensington eating anything out of the Thames, but London relied on the river for food early in its history. Fish bones have been found in Roman ruins in Southwark and sturgeon scutes have turned up in medieval remains at Westminster Abbey and Baynard's Castle.[8] Henry I, King of England from 1100 to 1135, supposedly died after overindulging in lampreys from the Thames. (This parasite fish with a mouth full of monstrous teeth is still a delicacy in parts of Europe and Asia.) More abundant were eel and twaite shad, while the most important fishery before the 1800s was smelt. Here I'll focus on salmon, the "king of fish," because it's very sensitive to low oxygen and because there are many reports about it in the Thames from the late 1700s through today. It has been said that salmon was so abundant in the preindustrial Thames that indentures stipulated apprentices were not to be fed the fish more than once a week. In his book *The Tidal Thames: The History of a River and its Fishes*, Alwyne Wheeler says the story is a myth. Regardless, the demise and return of salmon mirror the destruction of the Thames during the 19th century and its revitalization in the 20th.

Wheeler says the best year for the salmon fishery in the tidal Thames was 1816. Then with the start of the 1820s, salmon became rare; within a decade or two, the fish was no longer seen near London. The coronation of King George IV in 1821 went without salmon, although a day later two were caught at Bugsby's Hole. The last salmon caught at Boulter's Lock, where a fisherman and his son kept meticulous records, was in 1824. A candidate for the very last Thames salmon is the fish caught by William Yarrell in June 1833. Salmon were caught or seen further downstream of London throughout the 19th century, but these sightings were rare and newsworthy enough to be written up in sports magazines. In October 1970, an antique shop in London had on sale a stuffed salmon caught at Gravesend in May 1865. Maybe it was 1870; it doesn't matter. By then the king of fish was long gone.

The Thames lost its salmon and many other fish as London's population exploded from about a million at the start of the 1800s to over 6.5 million by the end of the 19th century.[9] To keep the engines of commerce humming, the Thames had to be improved to make it more navigable as well as to reduce flooding. To deepen its main channel, weirs were built bank to bank, interrupted only by pound locks to permit ships to pass. (The river today has 45 locks, each with one or two weirs and a lock keeper living nearby.[10]) Fishermen took advantage of the weirs to catch the fish stopped from migrating upstream. The weirs also slowed down the river flow, leading to the silting up of spawning grounds essential for the salmon life cycle—that's if the spawning grounds hadn't already been dredged out to further improve navigation.

Salmon and other migrating fish can slip past the locks when they open for ships, and not all of the spawning grounds were ruined. But another barrier couldn't be breached. That was the Thames's dead zone. The lack of oxygen downstream of London was more effective than any lock or weir in stopping fish from migrating to essential spawning grounds. The dead zone, or "oxygen sag," was likely kilometers long (Fig. 1.2) and undoubtedly extended from the bottom to the surface, in contrast to the hypoxic waters hugging just the lowest layer of the northern Gulf of Mexico and in other coastal dead

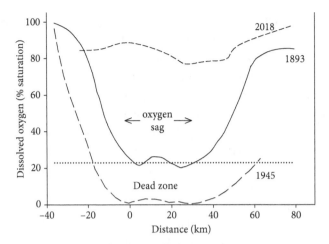

Figure 1.2 Dissolved oxygen levels in the River Thames in 1893, 1945, and 2018 in summer. The distance scale starts at London Bridge (zero kilometers); positive distance indicates locations downstream of the bridge. The dotted horizontal line is roughly equivalent to 2 mg per liter, which defines a dead zone. The 1893 and 1945 data are from L. B. Wood (1982). The 2018 data were provided by Mike Best, the Estuarine and Coastal Monitoring and Assessment Service, Environment Agency (UK).

zones. The Thames dead zone was too long and too thick for a fish to navigate before finding suitable spawning grounds. The anoxic and hypoxic water meant that even parts of the river with adequate oxygen could no longer be used as habitat for smelt, shad, and eel as well as salmon. Only some microbes could survive in the dead zone itself.

"Best of All Sanitary Reformers"

As the fishing records indicate, the River Thames was relatively clean in the early 19th century, and the Great Stink was still decades away in the future. The river was somewhat insulated from the wastes generated on its banks. London's early sewers were meant to carry surface water, not human waste, out of the city and into the Thames; until 1815, it was illegal to throw anything into the sewers. Londoners dumped wastes and garbage into alleys or the streets in spite of the laws against the practice, and they disposed of human feces and urine into cesspools. Some of these brick- or stone-lined pits were dug deep enough to reach a sand layer through which the sewage liquid seeped into the groundwater where some Londoners got their drinking water. The cesspools were irregularly cleaned out by night soil men who sold the human manure to farmers. In 1810, London had more than a million people and two hundred thousand cesspools,[11] and the Thames was still home to salmon. That would change as the cesspools were filled past their capacity by London's growing population. Cesspools were not the solution to handling the human wastes from a metropolis.

The cesspool problem was eventually solved by the toilet, or the water closet to use the British term favored in Stephen Halliday's book, *The Great Stink of London*. Halliday says the toilet didn't become common in London until 1810 even though Sir John Harington installed one in his house and another at the palace of Queen Elizabeth, Sir John's godmother, in the late 16th century. The first toilets, each using a lot of water ("water closet" easily earns its name), emptied into already overflowing cesspools. After the ban on connecting houses to sewers was lifted in 1815, toilets emptied into the sewage system that in turn discharged into the Thames. As cesspools became a thing of the past, a new problem arose. In 1840, reporting to the Select Committee on the Health of Town, Thomas Cubitt observed, "The Thames is now made a great cesspool instead of each person having one of his own."[12] Instead of overflowing cesspools, Londoners had a Thames overflowing with human excrement. The toilet and improvements in London's sewers efficiently transported raw human wastes from houses to the river.

Turning the Thames into a flowing cesspool didn't just ruin pleasure cruises on the river and disrupt meetings of politicians and bureaucrats in Parliament. It cost lives. It is difficult to imagine today that people would use water from such a polluted river, but the dangers of drinking contaminated water weren't clear in the early 1800s. According to one popular idea, pestilence came by air: miasma, or the vapor emitted from sewage and rotting garbage. Odors caused by the lack of oxygen didn't just turn your stomach, they made you sick and could even kill you. The miasma hypothesis seemed obvious, supported by the clear correlation between stench and diseases. One such disease was cholera, which killed nearly 40,000 Londoners during four epidemics in the 1800s.[13] By the mid-1800s, however, scientists began to realize that cholera was caused by something in drinking water. Many popular descriptions of this history highlight John Snow and his map showing the clustering of cholera victims around a water pump on Broad Street in Soho. The Broad Street pump had been contaminated by water draining from a nearby house where a mother had washed soiled diapers, or nappies in London English, from her child who had died of cholera. The oft-told story is that Snow removed the handle from the pump and stopped cholera from spreading. In fact, the disease was already declining by the time the pump was disabled.

More important was Snow's work on cholera spread by the Thames, London's main source of drinking water.[14] He took advantage of a natural experiment involving two water companies. During the 1848–1849 epidemic, both companies withdrew sewage-contaminated water from the river. By 1853 and the arrival of the next cholera epidemic, one company, Southwark and Vauxhall, still used sewage-contaminated water while the second company, Lambeth, had moved its intake pipes upstream where water was at the time relatively pristine. By going door-to-door to tally up the fatalities, Snow found water from Southwark and Vauxhall infected more people with cholera than did water from Lambeth. The work was important in disproving the miasma hypothesis and in showing that cholera was carried by water. The City of London ordered Southwark and Vauxhall to move its intake pipes upstream of the sewage outfalls.

Moving the intake pipes helped, but it wasn't the complete solution. As the city grew, more and more sewage was pushed upstream, once again threatening the water supply. And moving the intakes didn't stop the fetid odors wafting from the river, cumulating in the Great Stink. The next step was moving the sewage outfalls away from London, work that was overseen by the Chief Engineer for London's Metropolitan Board of Works, Joseph William Bazalgette.[15] Only 39 years old at the time, Bazalgette designed and oversaw construction of an elaborate and very expensive sewer system that diverted waste to the Thames estuary, east and downstream of the City. The plan was to hold the sewage until it could be released untreated via two huge outfalls along a three-kilometer stretch of Barking Reach. Work on the sewer system began at the end of January 1859, six months after the start of the Great Stink, and finished 15 years later in July 1874, with the opening of the Chelsea Embankment. By then London was mostly free of cholera.

Until this point in the Thames's history, dissolved oxygen levels were closely linked to two problems—the river's stench and diseases—which dominated the news of the day and demanded the attention of the highest levels of British government. (The dead-zone driven demise of the Thames fishing industry was much less troublesome. Londoners made much more money by other means and got their fish from other fishing grounds.) Something had to be done about both problems, especially the diseases that killed thousands. *The Times* pointed out in 1848 that "The cholera is the best of all sanitary reformers, it overlooks no mistake and pardons no oversight."[16] The miasma idea explicitly tied disease with odors caused by oxygen's absence. But then, the connection with oxygen was weakened as the miasma idea was discredited and the true cause of cholera became known. Low oxygen doesn't cause cholera, and it wasn't the reason why the Thames had the disease's causative agent, the bacterium *Vibrio cholerae*, which thrives with oxygen, although it can grow without the gas. The bacterium was in the Thames because the river was contaminated by feces from cholera-stricken people. With the new sewage outfalls at Barking, Londoners got safer drinking water, but the Thames lost its best advocate, cholera, for stopping the pollution and bringing back oxygen to the river.

Stopping Sewage's "Evil Effects"

Bazalgette's sewer system solved London's waste problem by relocating it downstream to Barking, but the people of Barking Reach weren't happy living and working near a sewage outfall. To make matters worse, the ebbing and flooding of tides caused the sewage to linger at Barking rather than to flow uninterrupted to the sea as it was supposed to. Even though the water was too salty to drink, Barking residents were still concerned about diseases carried with the sewage, and river pilots complained of headaches and nausea caused by malodorous water and mud. Barking used to be one of the largest fishing ports in England, although it had declined by the time it started receiving London's wastes.[17] The sewage outfalls sealed the town's fate. The outfalls became more widely known when the paddle steamer *Princess Alice* sank near one in 1878 with a loss of 650 lives. Survivors talked about the dreadful smell and taste of the water, and bodies recovered from the river were covered with a vile slime not found in clean water. An inquiry into the disaster considered whether some deaths were caused by the sewage or by hydrogen sulfide and other toxins released as the *Princess Alice* sank. The inquiry concluded that the cause of death was a lack of oxygen: asphyxiation by drowning. One investigator remarked, "I think I could swim and vomit at the same time," although he hadn't tried.[18]

As a result of the *Princess Alice* disaster, a Royal Commission on Metropolitan Sewage Disposal was appointed to determine if the sewage discharged into the Thames by London's Metropolitan Board of Works had any "evil effects."[19] If there were effects, the Commission was instructed to come up with a solution. It was 1882, the year William Joseph Dibdin became the chief analytical chemist for the Board after the death of his predecessor, T. W. Keates. Son of an artist and grandson of a poet,[20] Dibdin first followed his father's profession, but then he turned to chemistry and worked under Keates starting as a gas examiner in 1876. Within five years, Dibdin became Keates's principal assistant. Even before Keates's death, Dibdin was heavily involved in defending the Board against its many critics and was charged with fixing the sewage problem when he became the chief chemist.

A solution became necessary after the Commission decided that sewage did have evil effects and that it couldn't be dumped into the Thames untreated. The treatment devised by Dibdin was to precipitate the nasty stuff out of solution with lime and iron sulfate and then to dispose of the resulting solid sludge on land or out to sea. Precipitation facilities were built at Beckton and Crossness in 1887 to 1891.[21] These became small communities, complete with a school that doubled as a church on Sundays; any workman not at church had to explain himself to the superintendent. In addition to compulsory church attendance, what distinguished these facilities from others was the importance of science in maximizing the efficacy of the treatment.[22] Whereas facilities at other cities simply added the same amount of lime and iron sulfate, Dibdin's approach was to tailor the addition to the amount of organic material in the wastewater. Unlike elsewhere, he hired analytical chemists who used dissolved oxygen to estimate the amount of organic material in the sewage; today the method is called "biological oxygen demand," or BOD. Dibdin was a pioneer in wastewater treatment.

Another source of information Dibdin used for evaluating his sewage treatment approaches was dissolved oxygen levels in the Thames. In 1894 Dibdin published an analysis of 6400 samples from three years of data from Teddington to Nore, from the head of the tidal Thames to the river's mouth.[23] The data helped to show the Thames was improving. Dissolved oxygen levels went from dead zone levels (less than 20 percent saturation) in the summer of 1885 to over 50 percent of saturation ten years later.[24] The wastewater treatment efforts were paying off. Although conditions were not good enough for salmon, other fish returned. Dace and roach became plentiful above Putney where none had been seen for decades, and whitebait and flounder reappeared as the 19th century came to an end. The president of the British Institution of Civil Engineers expressed the hope that "it was not beyond the reach of possibility that Londoners might again see salmon ascending the Thames, as they did before that noble stream was so infamously polluted."[25]

Unfortunately, Londoners would have to wait nearly 100 years before seeing salmon in the Thames. The growth of the city in the early 1900s overwhelmed the treatment methods devised in the 1800s, and wastewater overwhelmed the Thames again. While dissolved oxygen remained okay through 1910, levels then plunged and the dead zone returned.

Before leaving 19th century London, Dibdin deserves a few more words. Other accounts of London's Great Stink highlight the accomplishments of people like Bazalgette who built London's massive sewer system or Edwin Chadwick, a social reformer who pushed for safe drinking water for London's poor. But it was Dibdin who arguably did the most for reviving the Thames in the late 1800s. Key to his success was oxygen, such as his invention of the BOD method and his use of dissolved oxygen data to evaluate the river's environmental health. After retiring from the Board in 1897, Dibdin continued to work on wastewater treatment and then became "one of the greatest authorities in the country on coal gas" according to his obituary.[26] He died in 1925 after being badly scalded in an accident, the shock too much for "one of his advanced years." (He was 74.) His obituary opined: "His death was as he would have wished it, that is, he died in harness, having been well and busy right up to the time of the accident and keeping cheerful and bright until the end."

"Natural Process of Purification"

Other cities in Europe and in the United States also suffered from problems caused by low oxygen levels in their rivers. Paris in the 19th century was inflicted by cholera epidemics and stench from the Seine River polluted by cesspools and sewers carrying human waste, horse manure, and trash. Eventually the city had to follow London's path and separate the intakes of drinking water from sewage outfalls.[27] The two largest cities in the United States around 1900 did something similar but with American twists. During the last decades of the 19th century, New York started to build an aqueduct that would ultimately carry water to the city from a pristine source 190 kilometers (120 miles) to the northwest. At the same time, construction was started on a sewer system that discharged into the ocean south of the city. Chicago got its drinking water from Lake Michigan in the 1800s, even though the city dumped its sewage into the Chicago River, which emptied into the lake. As Chicago grew, so did its wastes, and the sewage began to comingle with the city's drinking water. The solution effected in 1900 was to reverse the flow of the Chicago River. As a result, the river emptied into the Des Plaines River via the Chicago Sanitary and Ship Canal, eventually reaching the Illinois River, far from Chicago's drinking water source, Lake Michigan.

Chicago felt it didn't have to worry about its wastes contaminating downstream cities because the self-purifying power of rivers would take care of the problem. As pointed out by Charles F. Chandler, a chemistry professor at Columbia University and president of the New York Board of Health, "Although rivers are the natural sewers, and receive the drainage of towns and cities, the natural process of purification, in most cases, destroys the offensive bodies derived from sewage, and renders them harmless."[28] This purification process was evident even in the 19th-century Thames. Although Barking was ruined by being too near London's outfalls, Dibdin's 1893 data showed that dissolved oxygen was up to half of natural levels (50 percent saturation) about 30 kilometers downstream of the outfall, and it was near 100 percent after another 20 kilometers.[29]

The scientific basis for this "natural process of purification" was generally understood by the beginning of the 20th century. By then, it was known that dissolved oxygen and bacteria were key in cleaning up sewage. In 1857, Louis Pasteur showed living organisms were necessary to transform organic chemicals, such as turning sugars in grapes to alcohol in wine. A couple of decades later, while working on London's waste problem, Dibdin showed that both microorganisms and oxygen were necessary to degrade the organics in sewage.[30] Oxygen concentrations are low in sewer pipes, so not much degradation occurs there, evident by the stench of the stuff spewing out at outfalls. Once in oxygen-rich water, however, bacteria use oxygen to degrade the sewage organics to carbon dioxide and other harmless, odorless chemicals. Oxygen alone can degrade rotten-egg hydrogen sulfide to inoffensive sulfate, but degradation is faster when both bacteria and oxygen work together. Given enough oxygen and river miles, the natural process of purification works.

A problem arises when there isn't enough oxygen. More precisely, the input of organic material in sewage can exceed the amount of dissolved oxygen in receiving water, which is what happened to the Thames in the 19th century. Bacteria degraded the sewage organics, stripping the river of its oxygen at the outfall, leaving noxious chemicals to escape. Oxygen couldn't be replenished fast enough by diffusion from the atmosphere, which is a slow process, especially compared to the sewage flowing in from millions of Londoners. The brown Thames was too turbid and murky to allow any algal photosynthesis, the other main source of dissolved oxygen. However, as the river continues downstream, oxygen steadily diffuses into the river, allowing bacteria to finish off degrading the wastes. Far enough downstream from the sewage outfall, diffusion of atmospheric oxygen back into the river exceeds use of dissolved oxygen by organic-degrading bacteria, and oxygen returns to natural levels. The dead zone ends and life resumes.

The natural process of purification would not be complete without the destruction of those "offensive bodies:" the pathogenic bacteria, protozoa, and viruses. As the sewage is degraded and oxygen returns, the pathogens are killed or inactivated by a complex suite of biotic and abiotic forces. Well, usually. Most of the time, most of the pathogens are gone by the time dissolved oxygen returns to normal. But enough may remain. Here's a case where oxygen is an imperfect indicator of water quality, or at least of water safe to drink. Today we know that a swiftly flowing stream in the backcountry is full of oxygen, but a hiker can still get sick drinking that water. Sweet-smelling water can be dangerous. The tie between oxygen and potable water was further loosened at the end of the 19th century when engineers discovered how to make contaminated water safe for drinking.

Safe Drinking Water, Expanding Dead Zones

Sanitary engineers devised two methods to obtain safe drinking water from impure sources. Even before London realized it could prevent the spread of cholera by separating intakes of drinking water from sewage outfalls, one of its water companies, Chelsea Waterworks, and a few utilities in Europe, used filtration in the early 1800s mainly to improve the taste and appearance of the water.[31] Passing water through a sand filter also removes the cholera pathogen, although that wasn't appreciated at first. Later it was discovered that sand filters work because the pathogens are trapped and killed by natural, nonpathogenic bacteria, and other microbes that form a biofilm on the sand particles. This process works best when the biofilm is supplied with well-oxygenated water. However, filtration is not enough for a city relying on heavily contaminated water. A filter that removes 99 percent of all pathogens sounds great, but one percent of a very large number, the pathogens in contaminated water, slipping past the filter is still a large number. An only near-perfect filter could still allow thousands of pathogens to escape, enough to sicken people drinking the filtered water.

That necessitated the development of a second way to make water safe for drinking: killing the pathogens without necessarily removing them. After much trial and error, engineers settled on chlorine. As with many advances in water treatment, London was again the first to try chlorine in the 1850s to combat odors and cholera-causing miasma. Chlorine was occasionally used in the late 1800s and early 1900s in Europe and the United States. Its use picked up after engineers figured out how to produce concentrated chlorine gas and to deliver it safely and economically to water treatment plants. A study of 13 large US cities found that more were using chlorination than filtration by 1915.[32] By the middle of the 20th century, most treatment plants used a combination of filtration and chlorination to produce safe drinking water.

It is hard to overstate the significance of this achievement. Access to safe drinking water has saved more lives than most advances in modern medicine. In the United States, from 1900 to 1940, clean water accounted for half of the increase in life expectancy, three-quarters of the decrease in infant mortality, and two-thirds of the decrease in child mortality.[33] Cities became safer places to live when safe drinking water became readily available.

However, the capacity to turn foul water into something drinkable did not benefit the sources of that water: rivers, lakes, and reservoirs. With advances in making safe drinking water, waterways lost the best of sanitary reformers, water-borne diseases, which had forced governments to do something about stopping the pollution. Wastewater treatment improved in the early 20th century but not enough to keep up with the growth of cities, leading to the release of inadequately treated sewage into the nearest water body. The resulting odors and destruction of aquatic life weren't enough to goad governments into action. The Great Stink and its effect on Parliament aside, urbanites stayed away from rivers and could avoid a dead zone's stench; high-price condos and homes, trendy restaurants, and tree-lined bike paths are recent additions to river banks of large metropolitan areas. Dead zones wiped out fisheries, but that didn't bother cities. They got their money and power from other sources, and it was cheaper to import fish and shellfish than to spend millions on a wastewater-treatment plant. A pioneer in public health engineering did not see much benefit in treating wastewater:

> It is, therefore, both cheaper and more effective to purify the water, and to allow the sewage to be discharged, without treatment, so far as there are not other reasons for keeping it out of the rivers. . . . The water works man therefore must, and rightly should, accept a certain amount of sewage pollution in the river water, and make the best of it.[34]

In 1906, the US Supreme Court affirmed that cities could pollute away. The state of Missouri had filed a suit against the state of Illinois to stop Chicago from polluting the Mississippi River after the city reversed the flow of the Chicago River.[35] Writing on behalf of the court, Justice Oliver Wendell Holmes asked whether "the destiny of the great rivers is to be the sewers of the cities along their banks or to be protected against everything which threatens their purity." The court answered yes, rivers are sewers, and dismissed the case.

As sewage flowed from large cities into rivers in the 20th century, the result was a replay of what we saw for London in the 19th century: the disappearance of dissolved oxygen and the creation of new dead zones or the return of one in the case of the Thames. Construction of wastewater-treatment plants, already falling behind population growth, wasn't helped by two world wars and the Great Depression. The dead zone in the Thames in 1950 was probably as big as it was during the Great Stink of the 1850s, as the oxygen sag occupied over 50 kilometers of the river downstream from London's sewage outfalls.[36] The mid-20th-century Thames was far from being a noble stream, and there was no chance of salmon returning.

Similar to the Thames, dissolved oxygen also decreased in the Delaware River south of Philadelphia, the largest city on the river, during the first half of the 20th century. My colleague Jonathan Sharp has put together an amazing picture of oxygen in the Delaware from the late 1800s to the present.[37] Many of the recent numbers come from Jonathan's lab; but for the early years, he had to turn to other sources, some more accessible than others. No, not everything is online. Jonathan went through the paper reports from the Philadelphia Water Department dating from the 1920s and 1930s, and he guesstimated one number for 1880 based on the abundance of American shad, the "founding fish." Legend has it the fish kept George Washington's troops alive at Valley Forge during the American Revolutionary War. John

McPhee debunks the story in his 2002 book *The Founding Fish*. Even if it didn't help win the revolution, the fish was abundant in the late 1800s, indicating successful migration and spawning in the river north of Philadelphia, so dissolved oxygen must have been high at that time. Direct measurements show dissolved oxygen was still at natural levels in 1890 (Fig. 1.3). But then it started to decline precipitously, reaching dead zone levels in the 1940s and 1950s. The drop in dissolved oxygen mirrored the increase in the number of Philadelphians and the growth of the city's industries, which included oil refining, chemical industries, and shipbuilding at the Philadelphia Naval Shipyard.[38] The port at Philadelphia has always been one of the largest in the country, but it was especially busy during World War II, one reason among several why the mid-20th century was the worst era for the Delaware River.

At the start of the 20th century, the Delaware River lost the best of all sanitary reformers, waterborne diseases, in its defense against being used as a cesspool. Philadelphia didn't need a clean river in order to have safe drinking water; the city used a sand filter plant starting in 1899 and chlorination in 1913 to make water from the Delaware and the Schuylkill River safe for drinking.[39] Those treatments reduced deaths from typhoid fever by nearly 90 percent. Similar to London and New York, Philadelphia also built an extensive sewer system that dumped untreated sewage into the Delaware River downstream of its drinking water intakes. Upriver of Philadelphia, the city of

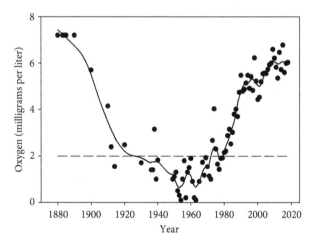

Figure 1.3 Dissolved oxygen over the years in the Delaware River south of Philadelphia. The dashed line indicates the oxygen level defining a dead zone. Data from J. H. Sharp and the Delaware River Basin Commission.

Trenton installed a sewage treatment plant in 1927, but the City of Brotherly Love did not build its first until 1951.

By then, the Delaware dead zone had become well-known, generating horror stories nearly as gruesome as London's Great Stink. Once abundant and the focus of several fishing companies along the Delaware estuary, shad plummeted in the early 1900s. Overfishing didn't help, but the river's dead zone ensured the fish's demise. Migration of shad as well as of eel and sturgeon was stopped by the oxygen sag, a dead zone that stretched from the bottom to the surface and from one bank to the other, for tens of kilometers. A 1946 article in the magazine *Colliers* called the Delaware a "slovenly river" and recounted a perhaps apocryphal story of naval pilots flying into the Philadelphia port area and smelling the fetid waters at 5,000 feet.[40] Hydrogen sulfide from the anoxic waters discolored buildings and ships, corroded instruments in factories, and sickened dockworkers. As with the Thames, oxygen levels tell the story of the Delaware's decline during the first half of the 20th century and how it hit rock bottom in the 1950s.

At about the same time, other water bodies in the United States and Europe reached a nadir. Three large cities—San Francisco, San Jose, and Oakland—dumped untreated sewage into San Francisco Bay during the first half of the 20th century.[41] Stripped of its dissolved oxygen, the Bay produced hydrogen sulfide that turned freshly painted buildings in San Jose grey and ruined rides across the new Bay Bridge. Dissolved oxygen in the Hudson River near Manhattan was at or near dead-zone levels, between 2 and 4 milligrams per liter, from 1920 to the late 1970s.[42] The Seine River was in bad shape as Paris outgrew its sewage treatment efforts during the 19th century. In the early 1880s, A. C. Géradin found that dissolved oxygen in the Seine was low, sometimes at hypoxic levels downstream of Paris, for more than 50 kilometers.[43] The record doesn't go back as far for the Scheldt River and estuary in Belgium, but we know untreated sewage caused oxygen to be low in the early 1970s.[44] The list of cities and their fouled waterways could go on.

Recovery of Urban Rivers in the 20th Century

Odors from sewage-contaminated water increasingly became a persuasive reformer; the smells had gotten intolerable, approaching Great Stink levels in European and US rivers. The stink wasn't the only reason raised against dumping untreated sewage into the most convenient waterway. Civil engineers came to recognize that sewage treatment would ease efforts to make safe drinking water, and it could reduce the foul taste not removed by filtration or chlorination. The middle of the 20th century also saw the rise of environmentalism and the realization that a river is a treasure to preserve, not a cesspool in which to dispose human waste and garbage. In *A Sand County Almanac*, first published in 1949, Aldo Leopold argued that rivers have a value other than to "turn turbines, float barges, and carry off sewage."[45] Even before the appearance of *Silent Spring* in 1962,[46] which inspired the environmental movement and rallied public opposition against pollution, Rachel Carson had published other highly acclaimed books about the environment in the 1940s, including *Under the Sea Wind*,[47] which came out as the United States entered World War II.

The end of the war allowed governments and society to turn their focus to the environment. In the United States, the first step was the Water Pollution Act of 1948, but the Delaware River and San Francisco Bay and elsewhere didn't improve until the Clean Water Act was passed in 1972.[48] The River Thames was helped by the Control of Pollution Act passed by the UK government in 1974,[49] while analogous regulation to protect the Seine went into effect in 1964 and 1992.[50] These laws forced cities to build and upgrade wastewater-treatment plants, made possible by help from central governments. In the United States since 1972, the federal government has paid out $650 billion for these plants, while cities have contributed another $163 billion.[51]

With more treatment plants, dissolved oxygen returned to large urban rivers. The Thames started to improve in the early 1960s when oxygen climbed out of dead-zone territory; in the last few years at London's northern outfall, concentrations have been typically about 80 percent of saturation in the summer.[52] Improvement came later to the Seine and has not been as extensive. Dissolved oxygen remained at dead-zone levels until the early 1990s and then jumped to about 50 percent of saturation around 1995; it's still only about 70 percent in the summer.[53] With respect to oxygen at least, San Francisco Bay is back to normal,[54] and the Hudson River at New York City is nearly so, even though Manhattan and Brooklyn continued to discharge raw sewage into the lower Hudson until the mid-1980s.[55] More success stories can be found for the River Mersey and the Firth of Forth in England, and the Rhine River and Elbe Estuary in Germany, and Boston Harbor and the Potomac River in the United States.

As oxygen returned to these rivers, so too did life, albeit with hiccups. The sightings of whales and seals in the River Thames made newspaper headlines, but the return of salmon may have caused a bigger stir. After an absence of over a century, a lone fish was caught in 1974, but it was enough to establish the Salmon Rehabilitation Scheme in 1979 to restock the king of fish in Thames tributaries.[56] Hundreds more salmon have been caught since the program started. New Yorkers and tourists have seen humpback whales, nearly 300 since 2010, frolicking at the mouth of the Hudson River;[57] while upstream near Hyde Park, Atlantic sturgeon cruise along the bottom looking for food and suitable spawning grounds.[58] These fish, which can be 16 feet long (nearly five meters) and weigh up to 800 pounds (over 360 kilograms), can now migrate up the river to spawn, unimpeded by low-oxygen waters. A recent study of a small river in Spain documented the less spectacular but more profound changes following the installation of wastewater-treatment plants in 2003.[59] As dissolved oxygen increased, so too did the diversity of benthic invertebrates, food for many fish in the water column. After being absent from the polluted river, brown trout have returned, as have Atlantic salmon, although abundances of most fish species are still lower than in other, more pristine rivers.

The Delaware River is another success story. The primary wastewater-treatment plants built in the 1950s for the Philadelphia area removed roughly half of the oxygen-depleting organic material, enough to set the river on a path of recovery. As the 1960s started, dissolved oxygen levels began to increase, reaching over 50 percent of saturation by the 1970s. Dissolved oxygen increased even more when secondary wastewater-treatment plants were constructed in the 1980s, costing cities along the Delaware over $1.5 billion.[60] When I first sailed on the Delaware in the late 1980s, dissolved oxygen was about two thirds of normal levels, while now it is closer to 80 percent on average. With the return of oxygen, many fish came back, including striped bass, white perch, and American eel.[61] Their populations have gone up and down over recent decades, and there are still problems, but at least dissolved oxygen isn't one of them.

The story is more complicated for the founding fish. Shad returned to the Delaware in the early 1970s when oxygen levels started to creep up, and for the first time in decades the fish was able to reach spawning grounds as far north as the Delaware Water Gap.[62] But shad numbers have trended downward since peaks in the 1980s. Although the gill-net fishery in coastal waters was closed in 2005, the fish hasn't bounced back. There is no shortage of possible reasons why. Some fish biologists think shad numbers are being kept down by striped bass, whose numbers increased when dissolved oxygen was restored to the Delaware. Striped bass are voracious, indiscriminate predators of many fish, including shad. Many other forces are preventing shad from completely recovering, including some we're behind. Even though the Delaware River famously has no dams, many of its tributaries with potential spawning grounds do. Dams are now being removed from the Brandywine River and other tributaries. The spawning grounds within the Delaware have been degraded over the years by onshore development and most recently by the US Army Corps of Engineers, which has been blasting rock outcrops to deepen the main channel of the river so that huge Panamax ships can pass. As exemplified by shad, the return of dissolved oxygen is an essential step but not the only one needed to make a habitat livable again and to ensure the complete recovery of aquatic life.

Uneven Progress, Other Problems

The Great Stink is no more, and cholera and typhoid fever are history—at least in rich countries. Dissolved oxygen has come back to rivers like the Thames and Delaware because governments had the money to build effective treatment plants. The same cannot be said for many poor regions of the world. On average, less-developed countries treat only about 8 percent of the wastewater they generate,[63] leading to the water quality and public health problems experienced by 19th-century Londoners. Dead zones in today's poor countries cause waterborne diseases when people have no choice but to drink contaminated water. A 2015 study estimated that each year cholera afflicts nearly 3 million people and kills 95,000 worldwide.[64] Countless others suffer from dysentery, which is especially debilitating for the malnourished. China hardly is poor today, but organic material from untreated or inadequately treated wastewater contributes to dead zone problems in the country's coastal waters like the Pearl River Estuary.[65] Less is known about problems in India. One study reported fish kills in the Tapi Estuary, about 300 kilometers to the north of Mumbai, after the estuary went anoxic due to the release of inadequately treated sewage from Surat City.[66]

Even developed countries have not completely solved their problems with wastewater, and even rivers with adequate dissolved oxygen are far from being pristine. The combined sewer systems, which are used especially in older cities to treat both rainwater runoff and domestic wastewater to-gether, are too often overwhelmed by heavy rainfall, leading to the release of fecal bacteria, toxins, and debris. Too many rivers are still polluted by hydrocarbons, heavy metals, plastics, endocrine disrupters, and industrial chemicals. Habitats for fish and other organisms are still being ruined by dredging and shoreline development, and fish ladders and elevators are im-perfect aids for negotiating past dams. But in many rivers, dissolved oxygen has returned, and hypoxia is no longer the main problem.

2

Dead Zones Discovered in Coastal Waters

Gene Turner wasn't supposed to be on the ship. He and Bob Allen had finagled their way onto the *R/V Oregon II* and sailed out of Pascagoula, Mississippi, in the spring of 1975 on the first of seven research cruises the two did over the next five years. They went on the cruises to measure dissolved oxygen in the Gulf of Mexico, though Turner and his oxygen work weren't part of the cruise plan for the *Oregon*. Built for the US Fish and Wildlife Service, the ship was designed to trawl for fish and shrimp, not oxygen, but Turner and Allen went along anyway. They had to sort fish to earn their keep, working four-hour shifts with overtime.[1] When not sorting fish, they could be found in a cook's closet that they had commandeered for their oxygen work. At the time, Turner knew oxygen better than fish, after working on the gas in Georgia coastal waters as part of his PhD dissertation, which he completed only months before the first cruise on the *Oregon*. His time in Georgia served him well. He learned about oxygen from his advisor, Larry Pomeroy, and about ecosystem science from a founder of the field, Eugene Odum. Years down the road, Turner's ecosystem perspective would prove to be key in explaining the Gulf's dead zone, equipping him to see connections between salt marshes of Louisiana, cornfields in Iowa, and nearly everything in between. But when the *R/V Oregon* left port in 1975, Turner was just interested in learning more about an important gas, dissolved oxygen, in the Gulf of Mexico.

Back then, not much was known about oxygen in the Gulf, especially near the Louisiana coast. Dissolved oxygen was measured in the early 1970s as part of an environmental assessment program to evaluate how drilling for oil and natural gas would affect the rich fisheries in the northern Gulf. Two years before Turner sailed, one study found that in the summer, oxygen in bottom waters south of Louisiana was near zero.[2] But this and other reports, at least the few sentences they had about oxygen, didn't attract much attention at the time. The reports focused on hydrocarbon and trace metal concentrations; the abundance of fish, mollusks, and polychaetes; and water movement at control sites and near oil rigs. Oxygen was just another thing to measure.

The reports didn't make a big deal of the low oxygen levels. It didn't help that they were published in venues far from the usual ones scientists follow. They were hard to get and not peer-reviewed, so even if unearthed, they were easily dismissed. So were the old stories from shrimpers about nets coming up empty because of "dead water." So, Turner didn't expect to find anything out of the ordinary when he sailed from Pascagoula in 1975.

What he found, however, was pretty alarming. In bottom waters of the northern Gulf in the summers of 1975 and 1976, oxygen was low, nearly a tenth of what it should have been.[3] Concentrations should have been about eight milligrams per liter, but in summer they were often lower than two, the typical cutoff defining hypoxia. Some levels were close to zero, indicating anoxia. Science had finally caught up with the shrimpers. Turner and Allen had stumbled on the Gulf's dead zone.

But Turner didn't know that in 1982 when he wrote up the results. He worked through explanations of how oxygen could be so low. He worried about their methods. Among all the samples with low concentrations, he focused on one sample that had high levels, supersaturated with oxygen, above that expected from just input from the atmosphere. He wondered if he just got the arithmetic wrong. When Turner first saw the low oxygen, he thought his homemade probe was broken. He would hold it up in the air, shake it to reset it, and then try again. It again gave the expected high levels in surface waters but the same unexpected low levels in deep waters. Also, his probe was backed up by another method, the venerable Winkler method, the gold standard for measuring dissolved oxygen. The probe was fine and the numbers were right. Oxygen is very low in the northern Gulf, dangerously low for most of marine life.

Turner then moved on to other explanations. Some were easily dismissed, while others are still debated today. He thought about all those oil rigs the R/V Oregon had to dodge while trawling for fish and shrimp, but quickly realized the oil companies couldn't be blamed this time. While oil seeping from leaking rigs or gushing from broken ones can be devastating, it couldn't explain Turner's low oxygen. Even the catastrophic Deepwater Horizon oil spill of 2010, the largest in the oceans so far, did not affect the size of the Gulf dead zone.[4] Turner wondered if the low-oxygen waters came from further offshore, from deep, hypoxic waters slurping into the shallows he and Bob Allen had sampled. In the early 1950s, F. A. Richards and A. C. Redfield, two giants in oceanography, found low oxygen in deep water of the Gulf's basin. But those deep, low-oxygen waters are too far away from the shallow, low-oxygen waters discovered by Turner. The deep waters also have more oxygen than Turner's oxygen-poor waters, so they were not the answer.

No, Turner knew that the low levels had to be due to high consumption of oxygen by bacteria and animals in the bottom water and sediments. Turner and Allen and everyone else knew that the breakdown of organic material can use up dissolved oxygen. That's what happened in the River Thames, the Delaware River, and many other rivers in the 19th and 20th centuries to create their dead zones. As in those rivers, bacteria and other organisms break down organic carbon back to carbon dioxide, using up oxygen in the process. You're doing the same metabolic reaction right now. You breathe in oxygen in order to break down the organic carbon in your last meal in order to gain energy. You oxidize or respire the organic carbon back to carbon dioxide, using oxygen along the way. You'll never run out of oxygen ("sucking all of the oxygen out of the room" is just a figure of speech) because the atmosphere has lots of the gas, but dissolved oxygen can be depleted by respiring organisms in aquatic habitats—when there is too much organic material. Diffusion of dissolved oxygen in water is too slow to counteract organic material degradation by bacteria. The question then became, where did that organic material in the Gulf come from?

One source Turner considered was the Mississippi River. It carries natural organics and potentially many other things from salt marshes, riverbanks, reservoirs, and soils of the thirty-two US states, about 40 percent of the continental United States, that make up the Mississippi watershed. Even two provinces of Canada contribute. There is a reason why the lower Mississippi is brown. But tea-colored tannins and other natural organics in the river aren't the main problem, or so Turner argued with turbidity data in his initial study. More convincing arguments came later. Soon we'll learn the Gulf dead zone is caused by what you can't see in the Mississippi River.

Turner settled on two other sources of the oxygen-consuming organic material. The most obvious one is sewage, the same stuff that caused London's Great Stink. West of the Mississippi's main channel, the Bayou Lafourche empties into the upper Gulf where oxygen concentrations were at their lowest according to Turner's data (Fig. 2.1). He pointed out the bayou serves as a "sewage conduit" for several southern Louisiana towns and told a reporter in 1982 that the problem was "all that garbage coming out of Bayou Lafourche."[5] The other source of the organic material Turner considered was more natural: algae, the plant-like microbes of aquatic systems. These microbes were abundant above the oxygen-poor bottom waters Turner sampled, although

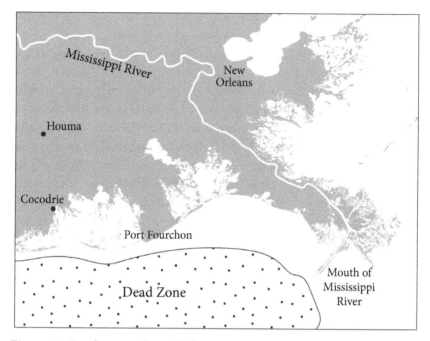

Figure 2.1 Dead zone in the Gulf of Mexico. An early hypothesis was that hypoxia is caused by sewage from the Bayou Lafourche with Port Fourchon at the mouth. That bayou and many others are not depicted in the map. Cocodrie is home to the Louisiana Universities Marine Consortium Marine Center and port for ships sailing to the Gulf's dead zone. A front-page article in the *Houma Daily Courier*, published in Houma 30 miles to the north of Cocodrie, was the first to use "dead zone" to describe hypoxic waters.

at the time, he didn't know why. Algae and the organic material they make supports all other life in oceans and other aquatic habitats, but when not eaten by herbivores, the unused material is like sewage. It sinks to the bottom where it is oxidized by bacteria back to carbon dioxide, using oxygen in the process. So much organic material sinks that oxygen runs out, turning the bottom layer hypoxic. The result is the dead zone.

Sewage or algae? If it's sewage, we know whom to blame. If the answer is algae, is their high growth in the Gulf somehow "natural"? What about before the 1970s? Has there always been a dead zone? Turner and Allen had three years of data, two years more than many studies, but still not enough to answer these and other questions. The answer didn't become clear for another ten years.

The Unlikely Environmentalist

Meanwhile, further north near Washington, DC, another dead zone was discovered soon after the initial work in the Gulf. Charles B. Officer led a team of estuarine ecologists in 1984 to examine the hypoxia problem in the Chesapeake Bay,[6] one of the largest estuaries in the world. Like most estuaries, the shores of the Bay and its tributaries are home to large cities, suburban sprawl, and many farms. As with many dead zones, there were early signs of a hypoxia problem in the Bay before alarm bells were sounded. A 1938 study by C. L. Newcome and W. A. Horne found low oxygen levels in the Chesapeake, but the low levels were interpreted as being normal, not a warning of serious problems. Officer and his collaborators didn't think the low levels were normal.

Officer was an unlikely environmentalist and champion for the Bay. He had made his fortune from his company, Alpine Geophysical Associates, that contracted with big oil companies searching for petroleum in the Gulf of Mexico or wherever oil could be found.[7] While the company's CEO, Officer ran in 1964 as a Democrat in a solidly Republican district to be one of New Hampshire's representatives to the US Congress. He lost by just a hundred votes or so, not helped enough by the coattails of Lyndon B. Johnson's landslide victory for the presidency that year. After having made "enough money at Alpine to care for my family's needs" and having "no desire to accumulate any more than was necessary," he sold his company and moved to Dartmouth College in New Hampshire. Dartmouth gave him a position but no office or salary. He didn't need the money or the space. He only wanted to write books, starting with one in his field, marine geophysics.

Like many American pioneers in oceanography, Officer's introduction to the marine world was thanks to the US Navy. After enlisting in 1944 when he was only 17, the Navy sent him to school, first to Williams College and then to Brown University where he studied physics. He got his PhD from Columbia University in geophysics, a field that was just emerging out of classified research conducted during World War II. Officer's first book written while at Dartmouth, one of nine he would eventually write, was a theoretical treatment of geophysics; his last book, published in 2009 with Jake Page, was *When the Planet Rages: Natural Disasters, Global Warming, and the Future of the Earth*. (He died in 2016 at the age of 89.) But the solitary life of a writer was not for him. He missed being listened to and having an impact. A friendship with John Ryther, who was at the Woods Hole Oceanographic Institution, got him back into research and thinking about environmental problems. The two became friends probably in Woods Hole, a beautiful village located on the elbow of Cape Cod, which Officer first visited in June 1949. Although he focused on aquaculture during the last part of his career, Ryther had published an influential paper in 1971 about nutrient limitation in the oceans, and he did some work on low-oxygen waters in the South Atlantic Ocean. Together Ryther and Officer wrote a paper in 1977 arguing that it was safe for the environment and cost-effective for coastal cities to dispose of their sewage in the ocean via outfalls.[8] The paper had lots about dissolved oxygen, so Officer was already thinking about the gas before turning to the Chesapeake Bay.

The Officer-led study of 1984 forcefully made the case that oxygen was disappearing from the Bay. His team's new data from 1980 were added to the old reports showing patches of hypoxia in the 1930s, 1949, 1950, and 1970. Officer argued that the volume of low-oxygen water in the Chesapeake had increased by five to six times from 1950 to 1980. In 1950, the first year when the Bay was completely surveyed, only a few parts of the mid-bay channel had low oxygen; by 1980, those waters had no oxygen, they were anoxic. Even worse, waters elsewhere in the Bay that once had normal oxygen were running dangerously low.

Although we now know the size of the Chesapeake dead zone is much smaller than the one in the Gulf of Mexico (the volume of the Bay's is about one-tenth of the Gulf's), hypoxia takes over a larger proportion of the Bay; bottom water hypoxia can make up as much as 20 percent of the Bay, but only about 3 percent on average of the Gulf shelf,[9] the shallow waters closest to shore where hypoxia occurs. As a result, hypoxia has an oversized impact on iconic Bay invertebrates and fish. Oysters, clams, blue crabs, and other bottom-dwelling invertebrates were known to be killed by hypoxia. Later work would show that low-oxygen waters also affect bottom-hugging or demersal fish, like flounder and striped bass,[10] all favorites of anglers and seafood lovers. Officer and his colleagues didn't use the term "dead zone," but that's what much of the Bay had become.

The Officer report doesn't say much about why low-oxygen waters had expanded so much in the Bay from 1950 to 1980. The report had a few words about increases in algal production and higher nutrient inputs, but that is all.

H. H. Seliger and colleagues had a very definite idea about the cause of the low oxygen they saw in the Chesapeake Bay.[11] Howard Harold Seliger had worked mostly on bioluminescence during his long career, but from his position at the Johns Hopkins University in Baltimore, perched at the northern end of the Chesapeake, he had also looked at red tides and understood the

Bay's physical oceanography, which would provide crucial for his oxygen study. Seliger documented "catastrophic" anoxia in the Chesapeake in 1984, just four years after the fieldwork by Officer and his collaborators. Strong words like "catastrophic" aren't used often in scientific journals, but this time arguably it was warranted. Seliger found that dissolved oxygen in bottom waters had decreased to zero, two months earlier than previously recorded, and more of the water column had low-oxygen levels than previously recorded. To make matters worse, in late August, they found hydrogen sulfide, the rotten-egg gas that stunk up London and the Delaware River. Some animals can shut down and survive without oxygen for days, but not many can withstand more than a whiff of hydrogen sulfide. Low oxygen defines a dead zone, but hydrogen sulfide makes it even deadlier.

The Seliger publication used "catastrophic" even when the full effects of low dissolved oxygen were not well understood. Now we know that some organisms are like the Eastern oyster (*Crassostrea virginica*), which can live for nearly a month without any oxygen by reducing their metabolic rate,[12] and other animals can switch their metabolism to a type (fermentation) that doesn't require oxygen, but even these animals eventually need oxygen. Even when not lethal, hypoxia changes behavior and metabolism and curtails growth and development.[13] It interferes with endocrine function and hormone production, and in the fish Atlantic croaker, hypoxia shrinks ovaries and testes, delays oocyte maturation, and slows down the swimming speed of sperm.[14] (In humans, hypoxemia caused by sleep apnea can dampen a man's sex drive.[15]) Hypoxia exacerbates diseases of marine invertebrates and fish. The eastern oyster becomes more susceptible to parasite infections, with effects lasting into the next year, and low oxygen may lead to outbreaks of mycobacteriosis in Chesapeake Bay striped bass[16] and papillomatosis in North Sea dab.[17] That low oxygen impairs vision was first reported by pilots flying at high altitudes during World War II. In the oceans, hypoxia is known to affect the vision of silver seabream,[18] and it can't be good for the many other fish, cephalopods, and arthropods that depend on visual acuity for finding food or avoiding predators. Hypoxia may modify genes and scar future generations even if oxygen returns to normal levels.[19] So, Seliger was right to call the problem "catastrophic." Less clear-cut was why dissolved oxygen had disappeared from the Chesapeake.

Unlike Officer, Seliger and colleagues had strong ideas about what caused the Bay's anoxia in 1984, and their hypothesis had little to do with organics and nutrients. They thought the problem was caused by the lack of mixing between the surface layer with lots of oxygen and the bottom layer with little. The two layers don't mix because the surface water is fresher or warmer or both, making it lighter than the deep water, which is saltier and colder. The light, oxygen-rich water sits on top of the dense, oxygen-poor water, like orange juice over grenadine syrup. This variation in water density, what scientists call "stratification," prevents oxygen from mixing into hypoxic bottom waters. The same physics applies to the Gulf of Mexico dead zone. In Louisiana coastal waters, stratification is caused by the Mississippi and the Atchafalaya rivers delivering a lot of fresh water to a very salty Gulf of Mexico. For the Chesapeake, the Susquehanna River supplies most of the fresh water that sets up stratification of the Bay. Officer didn't see any trend in the flow of the Susquehanna over the years he looked at, but Seliger told a different story.

Seliger identified several "climatic sequences" to explain the "anoxic catastrophe" in 1984. He noticed that in early spring of that year, the flow of streams in the Susquehanna River watershed was the highest ever recorded. Those fast-flowing streams contributed more water than usual to the Susquehanna, resulting in its flow being much higher than normal. All that fresh water coming into the salty Bay contributed to a light surface layer, strengthening stratification and further isolating bottom waters from exchange with oxygen-rich surface waters. The stratified water was also guaranteed to be undisturbed by wind because of unusual weather at the time, another part of Seliger's climatic sequence. The summer of 1984 was great for people living in the Bay region and not so great for Chesapeake organisms sensitive to low oxygen. A high-pressure zone persisted off the coast in August, so storms didn't disturb picnickers on land or sailors on the Bay. But storm winds also didn't break down the stratified water column or mix any oxygen-rich waters into the Bay's dead zone. All these physical factors, according to Seliger, explained the anoxic catastrophe of the Bay in 1984.

So, physics, specifically stratification, explains a dead zone, according to Seliger and his colleagues. Turner knew about stratification, but what he didn't know was the source of the organic material fueling oxygen consumption. Was it nutrient-enhanced algal production or from sewage? If asked, Officer and his group would have said something about nutrients and algae.

Figuring out the mechanism is the first step in figuring out how to fix the problem (Fig. 2.2). After the initial work in the Gulf and the old work in the Bay, it seemed possible that the dead zones in these waters were natural and were not caused by something humans were dumping into these waters. It

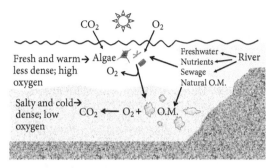

Figure 2.2 Formation of a dead zone by a river flowing into a sea or coastal ocean. The surface layer is less dense because of the freshwater brought in by the river. Oxygen is used up in the bottom layer when bacteria degrade organic material (O.M.) from algae and natural organics and sewage brought in by the river. Synthesis of organic material requires carbon dioxide (CO_2) and produces oxygen (O_2), while degradation is the reverse: it uses oxygen and produces carbon dioxide.

had been known since the late 19th century that some oceanic and lake waters are naturally hypoxic or anoxic. If physics explains dead zones, then perhaps we are off the hook. If hypoxia in coastal waters like the Gulf is caused by sewage, then we are to blame, but at least the answer is clear: don't dump raw poop into the most convenient water body. Algae-fueled oxygen use would be a bit harder to figure out because it leads to another question: why is algal production high in the Gulf of Mexico and Chesapeake Bay? Answering that question would be key to figuring out the origin of dead zones in these waters and elsewhere.

The First Dead Zone

While the Gulf and Chesapeake Bay dead zones were being discovered or rediscovered in the mid-1980s, Europe was realizing about the same time that it too had problems with low oxygen in its coastal waters. Europe has the dubious honor of being home to many of the biggest dead zones on the planet. One of the biggest is in the Baltic Sea.

The Baltic should have been the first region to be called a dead zone. Unlike the Gulf of Mexico, scientists have been studying the Baltic for over a century and a half. Its temperature and density were first measured back in 1771,[20] but work began in earnest a hundred years later when Gustaf Ekman "made some coasting expeditions" in 1877 to study the fishing grounds for herrings that had reappeared on the west coast of Sweden,[21] according to an account written by Otto Pettersson who had joined Ekman on the expedition. The sampling apparatus used by the two scientists was a product of 19th-century technology. Samples were taken with a rope-tethered bottle insulated by gutta-percha, a natural, rubber-like substance also used in early golf balls. Both ends of the bottle were open when lowered into the water and then were closed by an ingenious Rube Goldberg mechanism after reaching the desired depth. Ekman and Pettersson apparently didn't have much confidence in the bottle because they took water no deeper than 100 meters, where they were confident about which depth they had sampled. Only in waters shallower than 100 meters could they determine when the bottle closed from the "click" felt by a hand placed on the rope. Pettersson reported dissolved oxygen to be low in deep waters, a property he said is "highly characteristic" of water entering into the deep channels and basins of the Skagerrak and the Baltic. So, scientists knew of hypoxia in the bottom waters of the Baltic as early as 1890, yet there was no sound of alarm. It seemed normal, and probably it was.

Nearly one hundred years later, scientists were less sanguine. By 1969, enough data had accumulated over the years to reveal a low-oxygen problem in Baltic deep waters. The numbers showed a downward trend in dissolved oxygen over the decades, from about 3 milligrams per liter in 1890, to less than one milligram per liter in the late 1960s: hypoxia and a dead zone by anyone's definition. The author of a 1969 report, S. H. Fonselius, concluded "If the development continues, the Baltic deep water, the whole water mass below the halocline [where salt concentrations start to increase and the bottom layer begins] will soon turn into a lifeless 'oceanic desert.' "[22] He attributed the decline in oxygen to an increase in salinity and "waste deposal."[23] But the report didn't garner much attention. It appeared in an obscure journal with the uninspiring title, "Hydrography of the Baltic Deep Basins." Being ninety-seven pages long didn't help attract readers either; the Officer paper about the Chesapeake dead zone was five pages long. Plus, in the 1970s, the Baltic was facing even bigger problems than low oxygen.

Along with many environments before 1970, the Baltic Sea was contaminated by DDT, PCBs, mercury, and many other nasty chemicals. Oil spills were too common, and waste from pulp and paper mills was dumped into the Baltic too carelessly. A Swedish smelter released so much arsenic that it interfered with measuring phosphate, an essential nutrient. Fish stocks were running down, although that was thought to be due to overfishing, not pollution. The low-oxygen problem had to get in line behind those other pressing environmental concerns. By the mid-1980s, however, after DDT was banned and the other noxious chemicals declined, the Baltic had bounced back at least for some animals. Today the white-tailed eagle numbers over 4000, up from less than 700 during the DDT-filled days. Likewise, the great cormorant has rebounded to over 150,000 from a low of 4000 breeding pairs.[24] The banning of DDT along with new hunting regulations paved the way for the recovery of seals (ringed, grey, and harbor), although the harbor porpoise remains endangered.

The bad news is that hypoxia is today a bigger problem than ever, and not just because it doesn't have to share headlines with oil spills, DDT, and arsenic. The Baltic Sea now has the dubious honor of being home to the largest anthropogenic dead zone in the world, although there are other contenders, as we'll see. Today, it covers about 70,000 square kilometers, making it nearly five times bigger on average than the Gulf of Mexico's dead zone. The Baltic dead zone is the size of some European countries, not just a US state. Not only is it bigger, but the Baltic dead zone lasts longer—all year in fact, unlike the summertime visits by hypoxia to the Gulf and Chesapeake Bay. And to top things off, in some parts of the Baltic, oxygen drops down to low enough levels that hydrogen sulfide is produced.

The result has been the mass mortality of clams, mussels, and other benthic invertebrates, perhaps as much as 3 million tons;[25] in 2002, one hypoxia attack in the Danish straits killed off 300,000 tons of benthic fauna, about equal to the weight of all human Danes. These losses mean fewer demersal fish that depend on benthic invertebrates for food. Overfishing is the main problem with a popular fish, Baltic cod, but hypoxia doesn't help. Adults can escape to well-oxygenated surface waters, but when cod eggs sink to the hypoxic bottom layer because of their density, they fail to develop.

Algal blooms are twinned with all dead zones, if only as a possible source of organic material that drives oxygen depletion. Blooms are especially a problem in the Baltic Sea. In addition to algal blooms, the Baltic has experienced massive blooms of potentially toxic cyanobacteria that can cover up as much as 200,000 square kilometers,[26] much larger than England's 130,400 square kilometers. Algae and cyanobacteria share some similarities, such as some of their pigments (both have chlorophyll a), their microscopic size, and the way they carry out photosynthesis (the same as land plants). Cyanobacteria used to be called "blue-green algae" and sometimes still are.

But they are bacteria, not algae. Among several ways they differ from algae, some cyanobacteria are capable of turning nitrogen gas into useful forms of nitrogen nutrients via nitrogen fixation, ensuring that these microbes always have enough nitrogen. In spite of carrying out primary production like algae, cyanobacteria aren't as good as algae in feeding herbivores and in supporting the rest of the aquatic food chain. Cyanobacterial blooms have closed down beaches in the Baltic, and their toxins travel up the food chain to poison fish and shellfish.[27]

Cyanobacteria are a new addition to the list of suspects thought to cause a dead zone, but otherwise the list is the same: excessive organic material, physical stratification, or some combination of both. The possible sources of organics in the Baltic include sewage and algae along with cyanobacteria. Fonselius mentioned the increase in salinity, which would strengthen stratification differently in the Baltic than seen in the Gulf and Chesapeake Bay. In the latter two systems, stratification is fortified by less-dense fresh water from rivers flowing on top of dense salty water. Although the Baltic receives fresh water from the hundreds of rivers flowing into it, the occasional inflow of North Sea water also brings salt to the bottom layer of the Baltic, making it denser. The result is the same: more and stronger stratification. As with the two American dead zones, it was unclear which suspect was most important.

The Black Sea After the Deluge

The Baltic Sea dead zone is usually said to be the biggest in the world, if we consider only rivers and lakes, estuaries, coastal waters, and confined seas like the Baltic. We'll get to the open oceans later. Europe has another contender for the biggest-dead-zone title, the Black Sea. Deciding who gets the title depends on how you define "dead zone." The term usually appears in reports about unnaturally hypoxic or anoxic waters; but distinguishing natural from unnatural, meaning human-caused, can be tough, as we've seen already, and that's the case for the Black Sea. To understand oxygen in this confined body of water, it is useful to go back to its beginning.

The Black Sea started as a freshwater lake over 10,000 years ago. Its transformation into a salty sea was probably quite dramatic and perhaps epic-inspiring among the people on its shores. Unlike today, the nascent Black Sea once trickled west into the Aegean Sea and eventually into the Mediterranean Sea. Then about 7600 years ago, rising seas reversed the flow, and saltwater cascaded through the Bosporus, spilling over a rocky sill into the Black Sea with a roar and force of a Niagara Falls, or so say the more sensational interpretations.[28] Maybe the transformation took less than a year, but a more careful report says it was about 40 years,[29] still only a blink of a geologist's eye. The event has been said to inspire the Babylonian epic of Gilgamesh and the biblical story of Noah, although that's unlikely to be true.[30] In any case, today's Black Sea is still connected to the Aegean Sea only via the narrow straits of Bosporus, too narrow and too shallow to allow much oxygen-rich seawater to pass into the deep waters of the Black Sea (Fig. 2.3).

As a result, the bottom waters of the Black Sea are anoxic, even though the surface layer is full of oxygen, as in the Gulf of Mexico, Chesapeake Bay, and the Baltic Sea. From about 100 meters below the surface to the Black Sea's bottom, over 2000 meters deep in some places, there is little to no oxygen. This lack of oxygen makes the production of hydrogen sulfide possible. Some of this gas diffuses toward the surface where it is consumed by specialized bacteria, but most of it stays safely locked in the bottom layer, unless the Sea were hit by an asteroid big enough to release the hydrogen sulfide and set off a tsunami. If that were to happen, more people are expected to

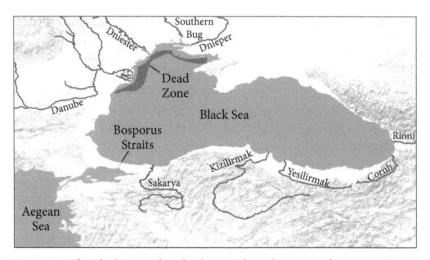

Figure 2.3 The Black Sea and its dead zone. The only source of oxygenated water for the deep layers of the Black Sea is the Aegean Sea via the narrow Bosporus Straits. The deep waters are naturally hypoxic or anoxic, but not so for the northwest coastal waters where the Danube, Dniester, and other rivers debouch.

die by hydrogen sulfide asphyxiation than by drowning.[31] Until the asteroid arrives, the Black Sea is the largest anoxic basin in the world, covering over 400,000 square kilometers, much larger than the Baltic dead zone with its 70,000 square kilometers. But the Black Sea doesn't get the largest dead zone title because the anoxia of the Black Sea's open deep waters is natural. Those deep waters probably have always been without oxygen since the Noah-like flood several millennia ago, long before enough people were around to do any damage.

What's not natural is the disappearance of oxygen from the northwestern shelf of the Black Sea where the water is less than 100 meters deep. Reports of hypoxia and even anoxia in this coastal region began to appear in the 1970s.[32] One event in 1975 was particularly devastating. Oxygen-poor waters "annihilated" benthic animals living where the Danube and Dniester Rivers empty into the Black Sea. Bottom-dwelling fish like goby and flounder were forced into the surf zone where they died, their corpses stinking up local beaches. Today the coastal Black Sea dead zone covers 40,000 square kilometers, larger than the Gulf of Mexico dead zone and second only to the Baltic's.[33] But perhaps the largest dead zone title should be awarded to the Black Sea in toto. The oxygen content has been decreasing even in the open Black Sea offshore from the northwestern shelf, and the depth at which hypoxia starts is inching toward the surface. So, the total volume of oxygen-depleted water in the Black Sea has been increasing. That loss of oxygen doesn't seem very natural.

The possible reasons, both natural and not so natural, why the Black Sea is losing oxygen are what we've seen already. The natural part is the now familiar combination of physics and biology. Fresh water from rivers like the Danube and Dniester creates stratification that cuts off the salty deep layer from the surface, sealing off poisonous hydrogen sulfide from escaping but also preventing oxygen from getting to the bottom water where it's needed. Oxygen is used up in the deep layer during the degradation of organic material. That's all natural and has been going on in the Black Sea since the Noah-like flood. The unnatural part is the large amount of oxygen-sucking organic material sinking to bottom waters of the northwest Black Sea coast. Possible sources for this organic material include the sewage coming down from the Danube and the Dniester, but also the Dnieper, the Don, and four other rivers that drain a third of Europe before flowing into the Black Sea. Those rivers also bring in nutrients that fuel algal growth, leaving behind organic material that sinks and depletes bottom waters of dissolved oxygen. It's the same list of suspects we've seen for the Gulf of Mexico, the Chesapeake Bay, and the Baltic Sea.

The number and size of dead zones have grown. In addition to the Baltic and Black Seas, Europe has the Adriatic Sea, the Scheldt estuary, and the German Bight. In addition to the Gulf of Mexico and the Chesapeake Bay, North America has the St. Lawrence estuary, Long Island Sound, and the Neuse River in North Carolina. Reports of hypoxia in Chinese coastal waters appeared only ten years ago, probably because no one looked before then. It would be remarkable if the Pearl River had any oxygen left after exiting Guangdong Province, home to over 100 million people. The East China Sea may be a contender for the largest dead-zone-in-the-world title.[34] A recent review pointed out that between 1960 and 2000, the number of dead zones doubled each decade; as of 2019, there were about 900 confirmed dead zones. Although the discovery of new dead zones seems to have slowed down, as possible sites in North America and Europe have been fully accounted for, other regions of the world have not been thoroughly explored. These regions include parts of Asia, the Indo-Pacific, and oceanic islands. Some are home to coral reefs, which may be threatened more by hypoxic waters than now appreciated.[35] It is possible to estimate the number of dead zones in these underexplored regions by looking at the number of people living in them; more people usually mean more hypoxia. Extrapolation from the population size–hypoxia relationship indicates that there may be as many as 1000 dead zones worldwide.

These dead zones have many things in common: sometimes sewage, always a supply of fresh water and nutrients, and lots of algae. The challenge is to figure out which of these things is most important. Answering that question will give us some ideas about how to solve the problem.

3

Coastal Dead Zones in the Past

Perhaps your sense of geography is like that of cruiseguy1016, a visitor to a website devoted to sightseeing cruises.[1] He asked about the distance to the Gulf of Mexico from New Orleans, the last big city on the Mississippi River before it empties into the Gulf. It was his first trip to the Big Easy, and he had seen a favorite cruise liner in port. Wondering how far the ship had to sail up the river to reach New Orleans, he went to a high point in the city, the I-10 bridge, where he expected to see blue water of the Gulf. Instead, he saw just more brown of the Mississippi. To answer cruiseguy1016's question, Host Clarea said that it took their ship about nine hours to traverse the nearly one hundred river miles from New Orleans to the Gulf. The journey would have been a few hours and miles shorter if the Mississippi River Gulf Outlet (MRGO) canal was open. But it closed in 2009 after the canal channeled the storm surge from Hurricane Katrina into the heart of New Orleans.[2] Even with MRGO, there would still be a 75-plus mile journey from New Orleans to the Gulf of Mexico.

That distance is relevant to thinking about one possible cause of the Gulf dead zone. Cruiseguy1016 might guess that the dead zone was because of pollution—the sewage coming out of New Orleans. He wouldn't be alone in thinking that. The idea occurred to ecologists when the hypoxia problem in the Gulf was first recognized in the late 1970s. One of the first publications about hypoxic waters in the Gulf mentioned sewage,[3] and the stuff has turned up in discussions of low-oxygen problems elsewhere. A review published in 1975 concluded that hypoxia in the Baltic was due to an "increasing load of organic material brought in by discharge of sewage outfalls and industrial wastes."[4] In Chapter 1, we saw the problems caused by untreated sewage dumped into rivers like the Thames and the Delaware, so Cruiseguy1016 would have good reason to blame sewage.

But the nearly one hundred miles separating New Orleans from the Gulf is one of several reasons why the good people of the Crescent City are not responsible for the Gulf dead zone. It's not the sewage from New Orleans or any city that's causing oxygen to be used up in bottom waters south of

Louisiana. As we saw for the Thames and other rivers rimmed by big cities, it is true that sewage organics can supercharge the use of oxygen downstream from a wastewater outfall, leaving behind hypoxia and anoxia. Fortunately, many of these problems have been solved by modern wastewater treatment, at least in rich countries. Along with several other cities in the United States, New Orleans installed its first wastewater-treatment plant in 1945.[5] But even if New Orleans were still dumping raw sewage into the Mississippi River, the stuff would not survive long enough to affect coastal waters. Bacteria would have enough time along the hundred-mile journey to degrade sewage organics in the Mississippi before they reached the Gulf. So, we can cross one suspect off the list of dead zone causes, at least for the Gulf. That sewage doesn't explain the Gulf dead zone was recognized soon after the initial reports of hypoxia appeared.[6] It's also not the main culprit behind today's dead zones in the Chesapeake Bay, nor in the Baltic or Black Seas. Here I'm talking about the turds, toilet paper, and other sundry organic material in sewage; we return to the nutrients in sewage in Chapter 5.

With sewage organics off the list, we're back to figuring out what causes dead zones like that in the Gulf. If sewage isn't the problem, perhaps the dead zone is just a natural occurrence, with natural causes. One way in which ecologists answer this type of question is to follow a habitat over time and look for correlations, in this case between potential dead-zone ingredients and dissolved-oxygen levels. Scientists call this type of data collection a "time series." Perhaps the best example of a time series, and certainly the most important one, is the program to measure carbon dioxide in the atmosphere. The program was started by Charles Keeling on Mauna Loa, Hawai'i, in 1958. Along with later programs established at several locations around the world, the Mauna Loa data have been crucial in exploring the rise in greenhouse gases and their impact on global climate. Time-series studies have also been instrumental in deconvoluting the possible reasons why an aquatic habitat goes hypoxic and whether hypoxia is natural for the Gulf. Were some regions of the Gulf always a dead zone? Or is hypoxia caused by something we are doing, which has gotten worse over the years? The early studies suspected we were to blame at least partially, but data were lacking. Ecologists had more data to implicate human activity as causing the Baltic Sea dead zone, but it wasn't clear when it started.

For the Gulf of Mexico, the early environmental assessment reports, published in the 1970s, had a few estimates of dissolved oxygen in the Gulf, and Gene Turner published his more extensive survey in 1982. But the time series for dissolved oxygen really began in 1985 by Don Boesch and Nancy Rabalais.

The Gulf of Mexico Time Series Begins

Boesch was primed to work on the Gulf's hypoxia problem by his days living near the Chesapeake Bay,[7] which had well-known problems with hypoxic waters. Perhaps being a native of New Orleans helped too. Boesch did his PhD at the College of William and Mary in 1971 and then took a position at the Virginia Institute of Marine Sciences on the shores of the York River, which eventually empties into the Chesapeake. There, he worked on benthic invertebrates as bioindicators—the aquatic equivalent of canaries in coal mines—of pollution in the Bay. These animals are also often the first dead-zone casualties. In 1980, he brought his bioindicator project to Cocodrie, Louisiana, and became the first director at the age of 34 of the Louisiana Universities Marine Consortium (LUMCON). You may remember Cocodrie was the town featured in the first newspaper article to use "dead zone." There wasn't much to LUMCON at first—just five trailers surrounded by salt marsh and bayous. Mike Dagg, the first scientist at LUMCON, said that when he arrived, he " . . . had nothing. I came down here from Long Island and for six months was living in one end of my trailer; my lab was on the other end. Thankfully, we did have electricity—most of the time."[8] LUMCON's building in Cocodrie is now comfortable and modern, topped by a cupola from which to survey marsh and bayous stretching to the horizon.

Boesch knew the shrimper stories about "dead water" where their nets turned up empty or with dead animals. Rather than helping out, however, he says "the shrimpers were kind of part of the problem, not part of the answer. They don't like to tell you things. They don't like government, they don't like science. . . . They were in large part in denial about hypoxia."[9] He also knew about the oil company environmental assessments and had heard of low oxygen in southwest Louisiana coastal waters where brine from the Strategic Petroleum Reserve was discharged. But getting money to look at the problem was difficult, evident from the fat file Boesch kept of rejection letters from state and federal agencies. Repeatedly measuring the same thing, even something as important as oxygen, was just monitoring work and not very innovative, according to reviewers. So, he used some of his bioindicator money for short trips on small boats to the eastern half of the Louisiana shelf close to Cocodrie. Eventually, things came together in 1985, the year when "dead zone" was first used by the *Houma Daily Courier*. It was the year he got a ship, the *R/V Pelican*, large enough to do long trips over the entire shelf, and with the help of US Senator John Breaux, secured money from the National Oceanic and Atmospheric Administration (NOAA) to begin regular surveys in Gulf shelf waters. Boesch then turned over the day-to-day operations of the NOAA program to Nancy Rabalais (Fig. 3.1).

Fiddler crabs would not seem the most obvious starting point for a career devoted to hypoxia, but that crustacean was the subject of Rabalais's PhD at the University of Texas (UT) in her native state.[10] She spent most of her time at UT's marine station in Port Aransas to do her PhD research. She had lived in Port Aransas before, when she had a job after college to check out benthic organisms, those marine coal mine canaries, as part of environmental assessments for oil and gas developments. After completing her PhD research in 1983, she moved to Cocodrie and continued her crustacean ecology and environmental assessment work. Her life changed dramatically in 1985. That was the year of her first *Pelican* cruise into the Gulf, far from the solid land of fiddler crabs. Then she met Gene Turner, that guy working in

Figure 3.1 Nancy Rabalais, hard at work on the *R/V Pelican* in 2007.

the cook's closet who published one of the first studies of low oxygen in the Gulf. At first the two were just friends and scientific collaborators, and then they became more than that. They were married in 1988. Getting married didn't stop Rabalais from doing science with her new husband. Rabalais says, "Our work on the Mississippi River and the hypoxia in the Gulf of Mexico is as a duo; Gene focusing upstream while I focus on the offshore end." Their names often appear together on publications, sometimes as Rabalais and Turner, other times Turner and Rabalais. However, she's the one once called "Queen of the Dead Zone."[11]

The annual surveys led by Rabalais have been instrumental in defining the hypoxia problem in the Gulf. The measurements since 1985 have revealed the huge, New Jersey-sized area covered by hypoxic waters. Rabalais and crew found that the dead zone moves around, shifting east and west along the Louisiana coast, and expanding or contracting north and south into the Gulf, from one kilometer to as much as 125 kilometers offshore.[12] (Other scientists have found hypoxic waters off other Gulf States, with Texas receiving the most attention, but these are separate, much smaller problems.) The depth where low-oxygen waters begin also varies. So, the thickness and volume of low-oxygen waters can change from year to year, even when the area does not. The Rabalais team usually does their annual survey in July, although recent work suggests that hypoxia is most intense mid-August.[13] Other surveys have found that bottom-water oxygen starts to disappear in May and remains low through August, sometimes as late as October in some years.

The extensive surveys have also been instrumental in figuring out why there is a dead zone in the Gulf of Mexico. The reason is the Mississippi River. There is a high correlation between river flow and oxygen levels in Gulf bottom waters. With river flow come all of the potential ingredients for a dead zone: fresh water, nutrients, and natural organic material. The drought of 1988 destroyed crops in the US Midwest, raised dust storms not seen since the Great Depression, and drained the Mississippi enough to expose century-old shipwrecks.[14] The dead zone during the 1988 drought was the smallest ever measured. The rain returned with a vengeance in 1993 and flooded the banks of the River, devastating an area greater than that inundated by the Great Mississippi Flood of 1927.[15] The flood of 2008 wasn't as bad, but the Bonnet Carré Spillway did need to be opened for the first time in 11 years so that overflow from the Mississippi River could take a shortcut to the Gulf via Lake Pontchartrain.[16] With those record flows in 1993 and 2008 came dead zones of record sizes.

Clearly, the area covered by low-oxygen waters has waxed and waned over the years with the flow and ebb of the Mississippi. Less clear-cut is whether there is any overall trend, up or down, since 1985. The short answer is it's complicated. Some publications report a significant increase in the area covered by hypoxic waters. But a more jaundiced view found that yes, dead-zone dimensions are going up with time, but only the upward trend in thickness is statistically significant so far.[17] One problem is that the amount of data is still small compared to the variation in dead zone measurements. Thirty-five years is a long time for a scientific study, but thirty-five isn't a lot of data points when oxygen levels change from day to day, with the seasons, and from one year to another.[18]

We're back to the questions raised at the start of this chapter: has the Gulf dead zone always been around? Could it even be natural? We know there were hypoxic waters in the 1970s, but what about before 1970? When did hypoxic waters first appear in the Gulf?

Sediment Time Capsules

To answer that, scientists needed to go back in time to estimate oxygen concentrations decades ago—or longer. The place for this type of time travel is at the bottom of the Gulf, in its sediments. The sediments contain time capsules that store relics reflecting past environmental conditions and a clock to tell geochemists the age of the relics. The top layer of sediments reflects conditions in the overlying water. If that water has low oxygen today, so too will the surficial sediments. Over time, the top layer becomes covered by new material settling down from the water column. What was once at the top now is pushed down deeper into the benthos. As a result, the relics in the top few millimeters of sediment are most recent while those deeper down are older. The depth of the sediment layer gives some indication of its age, but time is more faithfully recorded by the geochemical clock. Any stirring of the sediments by currents or burrowing animals screws up the relics, so not all spots on the bottom are amenable for time traveling. But when the conditions are right, the amount or number of an oxygen-dependent relic further and further down into the sediments tells us about oxygen concentrations further and further back in time, as recorded by the clock.

The geochemical clock is analogous to carbon-14 dating used by archaeologists to estimate the age of a corpse or a wooden crucifix. Although

geochemists also use carbon-14, the more common method for this application relies on lead-210 because of its shorter half-life. The principle is the
same for both isotopes. Both are unstable and thus radioactive, meaning they
spit out nuclear particles, gamma radiation in the case of lead-210, in order
to reach a more stable form. Lead-210 is produced naturally in the atmosphere and is deposited on the sediment surface along with everything else
raining down through the water column. Once cut off from its source in the
atmosphere, levels of lead-210 decrease as it decays to a stable form of lead,
lead-206. The relative amount of lead-210 in a sediment layer then reflects
the age of that layer. As time goes on, that layer will be pushed down deeper
into the benthos as more and more materials rain down. The result is that
deeper sediment layers are older with less lead-210.

Now that we know how to age our time capsule, we need to know about
relics that reflect past oxygen levels. One of the first relics, or proxies in this
field's argot, used by hypoxia studies was glauconite,[19] perhaps better known
as green sand. In addition to serving as a proxy for oxygen content, green
sand has been used as a pigment in oil paintings and as a soil amendment and
fertilizer. It's an iron-silicate mineral that forms when oxygen concentrations
are low. Pyrite or "fool's gold," another iron-containing mineral, this time
with sulfur, also appears in low-oxygen environments, and has been used in
dead-zone research. Other proxies include vanadium and forms of manganese. Although minerals are colorful and often informative, the most commonly used proxies for oxygen levels have been foraminifera, or "forams" to
their fans.

Forams are small, usually 1 to 100 millimeters, protozoa that live in the
water column and sediments everywhere in the oceans. The examples shown
in Figure 3.2 are from Tanzania.[20] While many different species are quite alive
and abundant today, it is the forams from the distant past that are studied
and useful in paleoceanography. Forams have been around since the Jurassic
Period when dinosaurs ruled about 200 to 145 million years ago. We know
that because forams left behind microfossils, their hardy shells, made of ce-
ment, or calcium carbonate. What is more, with a simple light microscope,
scientists can detect differences in shell shape and surface ornamentation—
characteristics that define the different foram species. Some forams are even
visible to the naked eye. The first recording of forams is believed to have been
in 5th century BCE by the Greek historian Herodotus who saw one type,
nummulites, in the limestone blocks making up the Egyptian pyramids. The
nummulites are quite big, at least relative to their microscopic cousins, about

Figure 3.2 Some foraminifera from the Eocene and Oligocene epochs recovered from Tanzania. The forams on the left are from the seafloor, and those on the right are from the water column. Their diameters are 0.20 to 0.75 millimeters. From P. N. Pearson (2012). Used with permission of the author and publisher.

1 to 5 centimeters (0.4 to nearly 2 inches) in diameter, and fossil nummulites up to 15 centimeters (6 inches) have been found in Turkey. These large forams get their name from the diminutive form of the Latin *nummulus*, meaning "little coins." Their round disk shape is not only just reminiscent of a coin; ancient Egyptians used nummulites as currency.

No microbe has been made more visible to the naked eye than forams. The tradition of visualizing forams started with Alcide d'Orbigny, the Frenchman who created the order of "Foraminifères" in 1826. He made hundreds of three-dimensional models of forams to publicize his work and to enable others to appreciate these creatures without a microscope.[21] The sculptures, about 5 centimeters in size or 40 to 200 times bigger than the original specimens, were made of limestone, and replicas were cast in plaster and packaged in wooden boxes for sale to devoted subscribers. I've heard that reproductions are still available, although I found only a framed print of two models, available for $119.99 (plus shipping and handling). The tradition of foram model building continued through the 20th century by many of the era's leading foram experts.[22] It reached its apogee with the Foraminifera Sculpture Park, which opened in 2009 in Zhongshan, China. The park consists of paths winding between sculptures several meters high or wide, modeled on microscopic forams usually seen only by specialists. The sculptures started out palm-sized, made for educational purposes by a marine biologist, Zheng Shouyi, at the Institute of Oceanology in Qingdao, China. She was encouraged to create the larger-than-life sculptures during a visit by an American marine geologist, Bilal Haq, then at the US National Science Foundation. Over five years under Dr. Zheng's direction, artisans created out of marble, granite, and sandstone 114 sculptures of over 100 different forams, representatives of life 250 million years ago to the present.

Scientists look at different species of forams to learn about the biology and chemistry of a habitat in the geological past. The stable isotope content of oxygen and carbon in foram shells tells paleoceanographers about carbon dioxide levels in ancient atmospheres and about the temperature and circulation of ancient seas. Another way to estimate temperature is from the ratio of magnesium to calcium, and the cadmium to calcium ratio says something about nutrient concentrations in the distant past. Oil companies look at forams to find out where oil and natural gas may be found. In theory, these companies could use other fossils, but forams are much more numerous and easier than dinosaur bones to sample using a drilling rig. Some species of forams prefer the turbid waters of river deltas and become fossilized in sandstone as plant residues turn into petroleum with the passing of eons. Forams help geologists to track down old, potentially oil-rich deltas as they shift around over geological time, such as the two-million-year-long migration of the Mississippi River delta from near the current border between Texas and Louisiana to its location today southeast of New Orleans.[23]

In addition to being a divining rod for black gold, forams can tell us about oxygen. The first studies found diversity and abundance of some forams to be low in the sediments under the hypoxic waters of today's northern Gulf of Mexico while other species were more abundant.[24] *Ammonia parkinsoniana* numbers decrease while *Elphidium excavatum* increase with low oxygen, and *Pseudononion atlanticum*, *Epistominella vitrea*, and *Buliminella morgana* all tolerate living in dead-zone sediments. Outside of the current dead zone, oxygen-needy species were more abundant than inside it. The simple correlations between current oxygen levels and these foram species gave rise to the A-E and PEB indices (the letters refer to the foram species names), tools for scientists to deduce oxygen levels in the past. By looking at these forams in sediment layers dated by the lead-210 clock, we can estimate a start date for the dead zone.

The initial reports indicated that the dead zone came into being sometime in the early to mid-20th century, with the exact year varying depending on the index, the foram species, and the sediment core. (A core is a cylinder of sediments a few centimeters in diameter by tens of centimeters to meters long, depending on location and luck during sampling.) In one sediment core, core G27 to be exact, collected in April 1989 by the Turner–Rabalais team, the A-E index for hypoxia started to increase between 1915 and 1925, while in another core (E30), the index didn't increase until after 1955.

The story became a bit more complicated, however, when longer cores were recovered, extending the record back more than 1000 years in some cases. In 2005, scientists at the US Geological Survey (USGS) recovered a single core, mellifluously labeled MRD05-04GC, which was over two meters long, more than twice longer than the cores collected previously.[25] Longer cores mean geologists can go further back in time.

It is clear that there were at least five peaks or "excursions," to use the paleoceanographic lingo, in the PEB index going down through sediment core MRD05-04GC and back into time. The peaks in the PEB index suggested that Gulf bottom waters had low oxygen as early as 300 years ago to over 900 years before the present. The PEB-based conclusions are backed up by two other oxygen-sensitive indices, types of metals whose abundance or chemical state changes with oxygen concentrations. Vanadium is high and a form of manganese (Mn IV) low when oxygen is high, and those two metals varied as expected with the PEB index over the 1000-year core. So, there are signs of the dead zone in the Gulf long before the 20th century when the Mississippi River drainage basin became covered by farms and cities.

The USGS scientists knew of the connection between the current dead zone in the northern Gulf of Mexico and the Mississippi River, so they wondered if the same connection explained the high PEB index values in the 19th century. So, they dug up past records, dating back to 1819, about the River's flow and paired that with their PEB data.[26] The old flow data came from Vicksburg, Mississippi, nearly 350 river miles north of New Orleans if you went by boat, a bit over 200 highway miles (330 kilometers) if you drove. Although far from the dead zone, it has the best historical record of Mississippi River flow. There are estimates even from 1863 when Vicksburg, a key city held by the Confederates during the American Civil War, was under siege by Union troops led by General Ulysses S. Grant. (Not until 1945 did the city officially but still tepidly celebrate the US Independence Day; Grant took over Vicksburg on July 4, 1863.)

The USGS scientists noticed really high flow, greater than two standard deviations higher than the average, in 1823, 1844, and in the early 1900s. Flow was one standard deviation higher during several years in the 1800s and 1900s. These years of high flow rates were the same, more or less, as the years with spikes in the PEB index and presumed low oxygen in the Gulf. So, not only was there low-oxygen water in the Gulf back hundreds of years ago, the intensity of that oxygen deprivation varied with output of the Mississippi River, just as it does today. Maybe low oxygen in the northern Gulf is the result of natural processes that have been happening since the Mississippi River started to flow into the Gulf millions of years ago. A closer look at the data, however, revealed some crucial differences.

Today's dead zone in the northern Gulf is not the same as the one in the distant past. The Gulf has seen bouts of low oxygen, perhaps even hypoxia, that came and went over the centuries. The PEB index in core MRD05-04GC suggests, however, that before the 20th century, low-oxygen waters occurred in intervals that lasted at most 25 years before disappearing, followed by the reappearance of normal oxygen levels that endured for decades. Today's dead zone, in contrast, arises annually without fail and has done so for the last 80 years at least. Also, when hypoxia did occur before the 20th century, it was limited to a smaller area close to the Mississippi River delta, based on evidence from a sediment core taken offshore.[27] This offshore core, long enough to capture a thousand years of oxygen-content history, showed no signs of low oxygen for centuries until about 50 years ago when the PEB index spiked up by 50 percent. Also, other cores indicate that the area affected by low-oxygen waters is much greater today than it was even in the early 1900s. Yes, some of the hypoxia in today's northern Gulf is natural, and there have been short periods of low-oxygen waters over the millennia. But something changed in the 20th century.

What changed is that the dead zone in the northern Gulf of Mexico grew into a behemoth, appearing without fail each year, starting around 1950. I'll use that year as shorthand to indicate the Gulf dead zone's birthdate, even though a single year gives a misleading impression of precision. Forams aren't perfect oxygen meters, and there are errors in the lead-210 clock and natural variability in the hypoxia recorded in the sediment cores. In any case, the sediment cores indicate that something happened midway through the 20th century that spawned the dead zone of today.

Dead Zones in Europe

Around the same time in some seas of Europe, oxygen also started to disappear. One of the first European dead zones to be documented was a region analogous to the Gulf of Mexico, the Adriatic Sea, that body of water separating the Italian peninsula from the Balkans (Fig. 3.3). It is fed by several rivers, including the Adriatic's version of the Mississippi, the Po River, which supplies about 70 percent of the fresh water going into the Adriatic.[28] The Po flows eastward across northern Italy for over 650 kilometers (400 miles) until it reaches the Adriatic south of Venice. The Adriatic Sea's connection

Figure 3.3 Adriatic Sea and the rivers flowing into it. The dead zone studied by Justić is in the northwest corner of the Adriatic where the Po, Brenta, and other rivers empty. The limited exchange between the Adriatic and the Mediterranean Sea is via the Strait of Otranto.

to the Ionian Sea and the rest of the Mediterranean Sea is by way of a tight, narrow gap, the Strait of Otranto, only 72 kilometers wide (45 miles), between the heel of Italy's boot on the west and the coast of Albania on the east. The Adriatic has had its share of environmental problems over the years.

The scientist leading the discovery of the Adriatic Sea dead zone,[29] Dubravko Justić, was just a PhD student at the University of Zagreb, then in Yugoslavia, now Croatia. Soon after his PhD, as Yugoslavia was disintegrating, Justić moved in 1991 to Louisiana State University with the help of a Fulbright Fellowship to collaborate with Gene Turner, Nancy Rabalais, and others on the hypoxia problem in the Gulf. He is now the Texaco Distinguished Professor in the Department of Oceanography and Coastal Sciences at LSU. One of Justić's current projects is to model hypoxia in the Gulf of Mexico using sophisticated computer programming. Gene Turner told me that Dubravko learned how to use computers by watching through a window into a room with the only modern computer that the University of Zagreb had at the time, then returning at night to practice what he saw.[30] He remains a good friend and colleague of Gene and Nancy.

Justić and colleagues knew about the algal blooms and other, more exotic environmental problems facing the Adriatic. These blooms may have occurred for centuries. Although blooms are crucial in supporting life in the oceans and all aquatic environments, blooms out of control can cause problems, including hypoxia. In the northern Adriatic Sea, blooms can contribute to the formation of blobs of mucus as big as dolphins, suspended a few meters below the surface like an unraveling mummy in a cheesy horror movie. Reports of these mucus blobs date back to 1729.[31] Whether because of the blobs or just nondescript organic debris sinking undramatically to the bottom, large numbers of benthic animals had died off in September 1980 and again in the same month of 1983, a few years before Justić began his PhD work.

To see if the die-offs could have been due to an Adriatic dead zone, Dubravko Justić rescued dissolved oxygen data from dusty file cabinets and obscure sources, and combed through reports dating back to 1911, written in German or Italian, published by the Instituto Sperimentale Talassografico and the Yugoslav Navy, among other institutions. Dubravko mentioned to me he became concerned as he worked backward in time.[32] After seeing little change in oxygen levels in the most recent data, he began to worry that he wouldn't have anything to say, not enough for a PhD dissertation anyway. He didn't have to worry. Once it was all assembled, the data told a simple story: dissolved oxygen had increased in surface waters since 1911. That may sound good, but it's actually a bad sign. It indicates high production of organic material by algae. Some of that organic material sinks to the bottom layer, fueling oxygen consumption there without any chance of replenishment from the atmosphere. So, the increase in surface oxygen meant that dissolved oxygen in the bottom waters was fated to decline. In fact, that's what the data showed. The northern Adriatic had become a dead zone.

The changes in oxygen were especially large midway through the 20th century. Dissolved oxygen significantly increased in the surface layer and decreased in the bottom layer after 1955. Justić found that the average oxygen level in the deep waters was not low enough to kill off large benthic invertebrates and to explain the massive die-off in the early 1980s, but the extremes were; some minima were low enough to be lethal. More recent studies have found that hypoxia continued to be a problem into the early 1990s, but then hypoxia became less frequent from 2004 to 2014 in some regions.[33] The coastal waters directly impacted by the Po River, however, continue to have low dissolved oxygen, a "strange color," and odor problems.

The main take-home message is the same we saw for the Gulf of Mexico; the mid-20th century seems to have been a critical time for the formation of a dead zone in the Adriatic. Going back further in time, studies using forams and other oxygen-sensitive indicators uncovered evidence of hypoxia in the 19th century and before.[34] But the forams point to signs of a dead zone appearing around 1930.[35] Regardless of the precise date, the northern Adriatic Sea had turned into a dead zone by the middle of the 20th century.

That was about the same time the Baltic Sea started to earn its title as the world's largest dead zone.

There were early signs of an oxygen problem in the Baltic, dating back to the gutta-percha insulated bottle days of Ekman and Pettersson in the late 1800s, but the timing and the extent of the problem were unknown. In the late 1960s, Fonselius worried about Baltic deep waters turning into a lifeless "oceanic desert,"[36] and Ragnar Elmgren,[37] then at Stockholm University, now retired, continued to sound the alarm in the 1980s. Over twenty years later, Daniel Conley and colleagues looked at over 300 sites scattered along the coast of the Baltic and found 115 of them had gone hypoxic from 1955 to 2009.[38] Although the frequency of hypoxia had increased since 1961, Conley didn't find any significant trend in the offshore waters. So, there were hints that hypoxia had become more common in the Baltic during the 20th century, but when that occurred wasn't clear at first, especially for offshore waters.

Those unknowns were addressed by a study led by Jacob Carstensen, based at Aarhus University in Denmark.[39] Carstensen and colleagues collated oxygen data collected over the years from two basins in the Baltic Sea: the Bornholm Basin, which is between Sweden and Poland, and the Gotland Basin, which is a larger and deeper part of the Baltic between Sweden and Lithuania. Sampling of the two basins has been nearly uninterrupted for over a century. Oxygen measurements were stopped by the two world wars, but water temperature went unmeasured only during World War I. Carstensen and colleagues noticed years when oxygen was low in the bottom waters of these deep basins in the early 20th century, but it always returned. Then midway through that century, things changed.

There was a large increase in the Baltic dead-zone area starting around 1950. The map in Carstensen's publication illustrating this increase is stark, with red patches indicating hypoxia and black for anoxia—the complete absence of dissolved oxygen. The red grows from small spots in 1906 to large, hemorrhaging splotches covering much of the Baltic by 2012. In the middle of the red zones are the black regions of anoxia. The black starts as small pinpricks on the map in the beginning of the 20th century, but by 2012, the map is covered in black, with anoxia swarming over nearly the entire Baltic. With more anoxia came more hydrogen sulfide, that rotten-egg gas, which kills faster than hypoxia. Other data show another disturbing aspect of the Baltic. Unlike the Gulf of Mexico and the Chesapeake Bay where hypoxia starts in spring and ends in late summer or early fall, oxygen levels in the bottom waters of the Baltic Sea are always low throughout the year. By all measures, midway through the 20th century, the Baltic Sea had turned into a huge, perennial dead zone.

Scientists doing time-series studies are continuing to explore the nuances of dead-zone formation, to evaluate the impact of hypoxia on aquatic organisms and ecosystems, and to look for signs of recovery. There is much we still don't know. However, a few conclusions are now clear. Before the 20th century, bottom waters of the northern Gulf of Mexico, the Adriatic Sea, and the Baltic Sea were sometimes depleted in oxygen, perhaps dipping to hypoxia levels. The same is undoubtedly true for many of the other current dead zones around the world. Hypoxia has come and gone over time, stretching back centuries (the extent of the Gulf data) if not millennia (the

Baltic Sea). These pre-20th-century hypoxia events indicate that nature has a hand in draining oxygen from bottom waters.

Then, dead zones around the world became commonplace in the middle of the 20th century. Once small and intermittent, the hypoxic zones in the Gulf of Mexico, the Adriatic Sea, the Baltic Sea, and many other dead zones have increased in size and frequency over the last 70 years. Although the exact year varies among locations, many dead zones expanded around 1950, and more came into existence. Recall that many of the 700 dead zones around the world were unknown prior to 1950.[40] This recent surge doesn't sound so natural. What are we doing to make these dead zones?

4

What Happened in 1950?

Green sand and forams in sediments and dissolved oxygen data in dusty filing cabinets told us that dead zones began to proliferate in coastal waters and bays in the 20th century. The year of 1950 is as good as any to mark the start of the dead-zone expansion, even though it may have been a few years earlier or later, depending the region. What is clear is that something happened midway through 20th century in the Gulf of Mexico, the Baltic Sea, and elsewhere. What exactly happened? Scientists have explored two hypotheses, the first involving a change in physics.

The nascent dead-zone regions could have become more stratified in the mid-1900s. More fresh water and higher temperatures in the surface layer or more salt and cold water at the bottom, or a combination of both, would make it harder for the two layers to mix. More stratification would reduce the transfer of dissolved oxygen from the surface to the bottom, giving bacteria more time to degrade organic material and use up the oxygen. Remember that the area covered by hypoxic bottom waters in the Gulf correlates with the flow rate of the Mississippi River. Droughts in the central United States mean less fresh water is delivered to the Gulf and the dead zone is smaller, while torrential rains cause high flow that delivers lots of fresh water to the Gulf and record hypoxia.

The second hypothesis is about the source of the organic material fueling oxygen use. When ecologists realized that dead zones in the Gulf and Baltic Sea were not caused by sewage organic material, they focused on another source of organics: algae. These microscopic, plant-like organisms are natural residents in all sunlit waters. Also natural is their occasional rapid growth and the accumulation of high biomass in algal blooms. These blooms feed crabs, shrimp, fish, and the rest of the food chain. What is not natural is excessive growth, stimulated by excessive nutrients: too much of a good thing, which leads to a dead zone. The algal blooms produce organic material that eventually sinks to bottom waters where it is degraded by bacteria, using up dissolved oxygen in the process. So, one hypothesis to explain the mid-20th-century proliferation of hypoxia is that the supply of nutrients to dead zones had increased, resulting in too much algae and organic material and a lot less oxygen in bottom waters.

Of course, both physics and biology are needed to make a dead zone. Physics is needed to prevent oxygen from being replaced. The algae have to be there to produce too much organic material. The question is, which is more important? More precisely, which caused the rise in dead zones in the 20th century? The answer informs our efforts to reverse the trend and bring oxygen and life back to aquatic habitats now devastated by hypoxia.

Let the Mississippi River Roll?

The Mississippi River today is not the same river that Mark Twain sailed on in the 19th century. Soon after New Orleans was founded in 1718 by Jean-Baptiste Le Moyne de Bienville, the French built small levees only a meter or two above the channel's top bank.[1] Over the years, the small levees begat bigger levees, eventually reaching 10 meters (over 30 feet) on average by the mid-1940s. Levees needed to grow because the bed of a river rises when its water and sediment are confined to the channel. By preventing floodwater from leaving the riverbed, levees speed water down the main river stem. Counteracting that acceleration, floodways and spillways, built to take pressure off levees, pause the flow south, as do locks and reservoirs. Starting in 1930, the US Army Corps of Engineers has dredged the main channel of the Mississippi River to maintain a navigation depth of three meters and lined the riverbanks with revetments to minimize erosion. Weirs and wing dikes redirect river flow. Many of these improvements were designed to control floods but may actually exacerbate flooding by the Mississippi.[2] Without the Corps and their political masters, the River naturally would take a serpentine route south, looping east then west, then back east as it makes its way to the Gulf of Mexico. But the Corps digs shortcuts like the Mississippi River Gulf Outlet canal for straighter paths and briefer journeys for barges and water. Although the Big Easy may let the good times roll, the Corps is much stricter with the Mississippi. With channelization and shortcuts, the River is now 200 kilometers shorter than it was in the 1700s.[3] Mark Twain saw this coming, sort of. In his memoir, *Life on the Mississippi*, published in 1883, Twain says "seven hundred and forty-two years from now the Lower Mississippi will be only a mile and three-quarters long, and Cairo [Illinois] and New Orleans will have joined their streets together."[4] (That statement is followed by Twain's real message: "There is something fascinating about science. One gets such wholesale returns of conjecture out of such a trifling investment of fact.") Given that the two cities are today separated by nearly 900 river miles, I doubt they'll share a mayor or city council anytime soon. But Twain was right about the transformation of Mississippi. With so much happening to the river, higher flow seems inevitable.

Changes in the watershed of the Mississippi River during the 20th century are more reasons to believe that river flow is faster now. Over the entire United States from 1901 to 2015, precipitation has increased by about 4 percent, with much variation around the country.[5] Less rain has fallen in the Southwest and Southeast, while more has doused the Northeast, Great Plains, and the Midwest. States lining the Mississippi River have seen some of the largest increases, 10 to 15 percent in Iowa, for example. That rain is falling on more impervious surfaces. The area in the Mississippi River basin covered by asphalt, concrete, roof shingles, and other hydrophobic material has increased by two and half times,[6] hastening rain water's journey to the river. With fewer chances of being soaked up by fields and forests, more rain seems destined to end up in the Mississippi and eventually in the Gulf of Mexico.

Yet in spite of all these changes, the Mississippi River flow has increased rather modestly, I think, and too late to explain the rise of the Gulf dead zone. In their 2003 study, the husband and wife team of Gene Turner and Nancy Rabalais saw no substantial change in river flow from the 1800s to 1990, and a study published a few years later saw about a 30 percent increase in river flow from 1940 to 2003.[7] That increase is about the same as what I see using the most recent data, but the change didn't start in 1940. I used an analysis called "segmented regression" to see when discharge started to increase (Fig. 4.1). The analysis says that discharge didn't trend up significantly until 1963 (plus or minus about 10 years), after which it has increased by about 6 percent per decade. That seems too late and too little to explain the rise of the Gulf dead zone in the 20th century.

Climate scientists have explored river flow in future worlds with different levels of change in climate and land use.[8] Depending on the scenario, flow of the Mississippi River could increase by up to 60 percent by 2090. The biggest increase would come in a future world with high carbon dioxide emissions and 4.5 degree Celsius warmer temperatures, which are some of the conditions envisioned in the A2 scenario, one of several alternative futures concocted by climate change scientists. Even in the environmentalist-favored

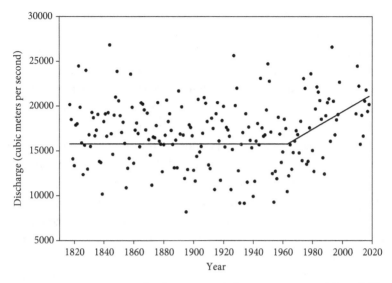

Figure 4.1 Mississippi River discharge at Vicksburg, Mississippi. The solid line is from a segmented regression analysis. Discharge varied up and down with no significant trend before 1963. After that year, it has increased significantly by about 6 percent per decade. Data from the US Army Corps of Engineers and the US Geological Survey.

B1 scenario, river discharge is projected to be about 15 percent higher than today.

But those are possible futures. The past changes in Mississippi flow cannot explain the rise in dead zones in the northern Gulf over the last hundred years. However, one big modification of the Mississippi has moved where hypoxia occurs in the Gulf.[9]

The story starts in 1831 when Captain Henry M. Shreve, an illustrious steamboat man and entrepreneur, became too impatient with Turnbull's Bend. Located about 70 river miles north of Baton Rouge, the Bend was a westward oxbow loop of the Mississippi that added 20 miles and several hours to the trip south. It had the Red River coming in on the north when the Mississippi zigged west and the Atchafalaya River leaving it on the south when it zagged back east. Shreve's solution was to dig a canal, Shreve's Cut, to bypass the Bend, allowing the Mississippi to flow straight south without the zig and the zag. The northern part of what remained of Turnbull's Bend silted up, but the southern remnant, eventually known as the Old River, fed by the Red River, kept flowing east, carrying water back into the Mississippi.

That was the easier route because the alternative, flow southwest into the Atchafalaya, was restricted at the time by a 30-mile-long logjam. The Old River began to get confused in 1839, however, when work started to clear away the logs. By 1840, the logjam was gone. The water continued to flow east into the Mississippi but only when the Red River was high; when the Mississippi was high, it went west into the Atchafalaya River, now the easier, shorter route (142 miles vs. 315 miles) to the Gulf for water from the upper Mississippi to take. Over the years, more and more water flowed into the Atchafalaya River as it became wider and deeper, such that by 1953, the Corps realized that the Atchafalaya River would take over, leaving no water for the Mississippi downstream of Shreve's Cut. As John McPhee melodramatically put it, "It was at Old River that the United States was going to lose its status among the world's trading nations. It was at Old River that New Orleans would be lost, Baton Rouge would be lost."[10] Although the cities wouldn't be "lost," their ports would be, and commerce down the Mississippi, which bears a large fraction of US exports, would be seriously threatened.

The solution was the Old River Control Structure, which grew into a Complex, a massive system of sills, floodgates, a navigation lock, and a hydroelectric station. Begun in 1955, the first couple of phases were completed by 1962, with more components added in the 1970s and mid-1980s. The goal of the Control Complex is to ensure that only 30 percent of the water from the Red River and the upper Mississippi River flows into the Atchafalaya River. The remaining 70 percent continues down the Mississippi to keep Baton Rouge and New Orleans alive.

The water dealt out by the Old River Control Complex affects the location of hypoxia in the northern Gulf. By feeding water to the Atchafalaya River rather than to the Mississippi, the Control Complex sends hypoxia-causing fresh water further to the west toward Texas. The mouth of the Atchafalaya River at Morgan City is about 250 kilometers by boat west of the bird foot delta of the Mississippi. The impact of the Atchafalaya is even greater than implied by the current 30-70 split between the two rivers. Whereas all of Atchafalaya's water empties out into shallow shelf waters most susceptible to hypoxia, only roughly half of the Mississippi River does. The rest goes south and east into the Gulf where hypoxia is less of a problem. As a result, in spite of the 30-70 split, the Atchafalaya and the Mississippi contribute about equally to the dead zone of the northern Gulf.[11]

However, the Old River Control Complex can't be blamed for the expansion of hypoxia in the northern Gulf over the last seventy years. The timing isn't right—it came too late, well after the dead zone became common and extensive. The Atchafalaya versus Mississippi tussle, without a change in total flow, can explain only the location of hypoxic waters, not the dramatic increase in duration and size of the dead zone in the 20th century.

Changes on Land and in the Mississippi River

So, what has changed over the last seven decades? Why has the dead zone grown over the years? We just saw that physical changes in the Mississippi and stratification aren't the answer. That leaves the other possibility: nutrients. During the 20th century, the Mississippi and Atchafalaya Rivers were bringing to the Gulf more and more nutrients that stimulated algal growth and eventually oxygen depletion. This increase in nutrients is called "eutrophication," sometimes "cultural eutrophication," to distinguish it from nutrient inputs by natural mechanisms. The role of nutrients was recognized in the late 1980s by scientists working in the Gulf and other locations with dead zones. Scientists long suspected that nutrients in rivers feeding into dead zones had increased over the years, resulting in the expansion of hypoxic waters in the 20th century. The challenge was to get the data to support their hypothesis.

The data came from time-series studies, showing how nutrients and algal growth have increased over time and how these changes in the water can be linked to what is happening on land.

Gene Turner and Nancy Rabalais used time-series data to explore how one important plant nutrient, nitrate, varied over the years in the lower Mississippi River. They found that nitrate concentrations had increased by four times during the years they examined, from the early 1930s to 1990. Gene and Nancy ended their study in 1990, but that year also was near the end of the large increase in nutrient levels. They did have a handful of nutrient data before 1950, from around 1905, but not enough to say much. To explore what happened further back in time, Turner and Rabalais turned again to the sediment time capsules.[12] They used the lead-210 method to date the sediments, as described before, and silica as a proxy for algal production in the past. Silica follows only one class of algae, the diatoms, but diatoms are the most abundant class when nutrient concentrations are high. Diatoms are important in supporting the aquatic food chain. If diatom growth is high, so too is the growth of fish and other aquatic animals. Diatoms can be followed by silica, the main component of glass, used to make diatom cell walls, essentially encasing the content of their cells within glass houses, which are exquisitely decorated with pores for exchange with the outside world. Because silica is resistant against degradation, the amount of it preserved in sediments reflects the level of algal production in the past.

Turner and Rabalais found that the amount of diatom silica in Gulf sediments was about the same from 1700 to 1800, and then it started to rise.[13] They connected this 19th-century rise in algal production to settlement of the Midwest by Europeans and the intensification of agriculture in the Mississippi River basin. Forests were cut down and the wood used for industry and construction and to heat homes. The tough sod of grasslands was broken up by professional prairie breakers to make way for fields of wheat and corn. Livestock were allowed to forage and to roam and root freely, digging and disturbing soil in unfarmed land almost as effectively as a plow. A visitor to the region observed, "There is no portion of the globe that is being exhausted of its fertility by injudicious cultivation, so rapidly as the Mississippi Valley."[14] That was in 1849. Later studies showed that erosion of cropland soil is much higher than that of soil covered by prairie grass or forests. The transformation of the wilderness to the Midwest's fruited plains resulted in nutrients washing into the Mississippi.

The 20th century would see even larger changes. While there was hypoxia during the 19th century, the afflicted area was probably smaller than now, and the dead zone did not reoccur every year. Sediment silica and by implication algal production increased dramatically starting in the late 1960s, coinciding with higher and higher nitrate concentrations. Levels of another important nutrient, phosphate, also increased, although the data are not as extensive as for nitrate. Those high concentrations have led to more algal production, more organic material, and less oxygen in bottom waters of the Gulf. The result is today's dead zone.

"The Tyranny of Oceanographers"

Here is what we know so far. The dead zone in the Gulf is formed by both physics and biology. The physics part is the delivery of fresh water to the Gulf by the Mississippi and Atchafalaya Rivers. Once in the Gulf, the fresh water along with warm temperatures stratify the water column and prevent dissolved oxygen at the surface from being mixed to the bottom. The biology includes production of organic material by algae and degradation of those organics by bacteria. Without new inputs, oxygen is depleted in the bottom layer as bacteria degrade organic material made by algae in the surface layer. Midway through the 20th century, levels of nutrients like nitrate and phosphate brought in by the rivers soared, which stimulates algal growth and production of organic material. When that organic material is degraded, dissolved oxygen is used up, and a dead zone is created. That's the standard story for the origin of the dead zone in the Gulf and in many other waters as well. There are, however, a few complications.

First, I need to say something about another nutrient, silicate, used by only one type of algae, diatoms, for their silica-laden cell walls. Diatoms are worth the special attention due to their importance in supporting the food chain, all the way up to fish. The complication is that, unlike the other nutrients, silicate levels have decreased by half since the early 1900s[15] because silica-rich sediments are increasingly trapped upriver before they reach the Gulf. The drop in silicate favors other algae and diatoms with thinner silica walls, which in turn effect changes in their predators and so on up the food chain. There is even concern that harmful algal blooms will become more common as silicate declines.[16] While important, I don't think lower silicate levels change our understanding of why the dead zone expanded in the Gulf mid-20th century.

Another complication is that there are more physical drivers shaping the Gulf dead zone than just the Mississippi and Atchafalaya Rivers. The strength and direction of the winds figure into its size and location. A strong westerly wind that blows long enough shrinks the hypoxic area, because it pushes the river plume to the east and offshore over deeper waters of the Gulf.[17] Winds, along with cooling temperatures from the passing of a cold front, break down stratification in autumn. Without stratification, oxygen-rich surface waters can mix with the hypoxic bottom waters—they are "ventilated," to use the oceanographer's term of art. That spells the end of the dead zone for the year. Hurricanes are the ultimate mix master. The dead zone in 2019 was expected to be huge after heavy rains in spring that caused historic flooding in the Midwest and Mississippi River basin.[18] Then Hurricane Barry stirred things up mid-summer, making the dead zone smaller than expected. But even a hurricane's impact may be erased if there is enough time for stratification to reset and for bacteria to use up oxygen in bottom waters. Hurricane Katrina, the monster that nearly destroyed New Orleans in 2015, hit the Louisiana coast in late August after disrupting hypoxia in coastal waters, but then the dead zone returned a few weeks later.[19] Winds and hurricanes complicate dead-zone physics, but again they don't really change the standard story about the origin of the Gulf dead zone.

Even if the main ideas about dead-zone formation don't need to be revised, information about silicate levels, winds, and hurricanes is important. It gives a more complete understanding of the extent and timing of hypoxia as set by nutrients and freshwater flow, according to the standard story. The final complication, however, could substantially change the story.

Rather than from algae, organics could come from the vast wetlands ringing the northern Gulf. Tom Bianchi, an organic geochemist now at the University of Florida, has championed the wetland organics hypothesis. I first met Tom at Tulane University in New Orleans when I gave a seminar there in 2004, a year before Katrina hit, forcing him and many others to leave the city.[20] He never went back. Eventually he found a new home in the Department of Oceanography at Texas A&M University where a colleague got Tom involved in work on the Gulf dead-zone problem. By this time, Tom was a tenured, full professor at Texas and settled enough in his career to take on controversy. He hadn't worked directly on hypoxia while at Tulane, although he knew the scientists at Louisiana State University and LUMCON in Cocodrie, about an hour and forty-minute drive from New Orleans. Bianchi usually focuses on deciphering the source and fate of natural organic material as it moves from large rivers into coastal waters. But one of Tom's studies did look at the hypoxia problem.

It combined data on stable isotopes and chemical signatures into a simple model to explore the source of organic material found in the northern Gulf.[21] The results suggest wetlands can't be ignored. As much as 39 percent of the organics in the Gulf dead zone region could come from wetlands in spring when the dead zone starts to form. Bianchi believes that these organics could be delivered to the hypoxic bottom layer by mobile muds coming from the Atchafalaya River. In addition to the natural delivery by rivers and from salt marshes, organic material could come from the very unnatural loss of wetlands in Gulf States, especially Louisiana. The state has lost about 5000 square kilometers of wetlands from 1932 to 2016, equivalent to the area of Delaware, roughly a football field every hour.[22] Although we don't have enough data from the early 1900s to be sure, a case could be made that sometime mid-20th century Louisiana wetlands started to disappear just as the Gulf dead zone started to expand.

Wetland organics definitely aren't part of the dead zone origin story promoted by Rabalais and Turner, so they responded with their own numbers.

Their data indicate that wetland organics aren't that important.[23] Yes, some of these organic chemicals make it to the northern Gulf and potentially support consumption of dissolved oxygen, but the amount is small compared to the organic material from algae.

Rather than get deeper into the details of the Rabalais–Turner rebuttal, I'll make two points. First, even Tom Bianchi would concede that algal organics are easier than wetland organics for bacteria to degrade; even if the wetland material is abundant, it may not fuel much bacterial degradation and oxygen use. The second point is based on a study that looked at how much of the waxing and waning of the Gulf of Mexico dead zone could be explained by stratification and nutrients like nitrate.[24] Like most, the study used nutrient levels in spring to explain hypoxia in summer because a couple of months are needed for algal uptake of the nutrients, production of organic material, sinking of that organic material to the bottom layer, and its degradation. The study found that nutrients were more important in some years, while in other years stratification was the winner. But overall, nutrient levels were a bit more important, 51 percent versus 38 percent for stratification, in contributing to Gulf dead-zone formation. Together, I calculate that the two factors explain nearly 90 percent of the variation in dead-zone area. That doesn't leave much room for wetland organics.

The back and forth between Tom Bianchi and the Rabalais–Turner team would have been just another disagreement about an arcane scientific point if the sparring had been confined to discussing obscure data in specialty journals. It's part of doing science, although the animosity coloring the debate is unusual and less understandable to an outsider. What elevated the fight to a higher level was the publication of a couple of provocative essays by Bianchi that even a politician could understand. The essays were published in *Eos*, far from an obscure specialty journal. It is the newsletter for the American Geophysical Union, one of the largest scientific organizations in the world. In one essay, Bianchi takes issue with the main Gulf dead zone origin story and the focus on nutrients as the way forward for solving the hypoxia problem. He and his coauthors said, "A management strategy based on only one aspect of hypoxia is likely doomed to fail."[25] A couple years later, another Bianchi essay said, "However, until scientific understanding of the quantitative relationships between nutrient inputs and their ecosystem consequences improves significantly, mandating a specific nutrient reduction target level is difficult to defend."[26] The essays had other criticisms.

I think I understand why Tom wrote those articles. He felt that wetland organics and other issues were being ignored, even repressed, because they didn't fit the standard story. He thinks that the loss of wetlands is a bigger, more immediate problem for Louisiana than the hypoxic waters sitting off its coast. He makes some good points. But Don Boesch says the *Eos* articles damaged efforts to shrink hypoxia in the Gulf.[27] We first met Boesch when he was the head of the lab at Cocodrie and was starting the Gulf dead zone research program, which Nancy Rabalais would eventually take over. By the time Bianchi's *Eos* articles were published, Boesch had become a heavy hitter in environmental politics as the president of the University of Maryland Center for Environmental Science. He said that Bianchi's *Eos* articles gave Midwest legislators more ammunition to fight efforts to minimize nutrient loss from farms. Why set regulations when scientists don't completely understand the problem? Don says the naysayers are helped by a "tyranny of oceanographers" who argue that there is more to hypoxia than just nutrients, that more research is needed to understand a complicated problem. He says that the legislators wanted to resolve issues like those raised in the *Eos* articles before spending any money on solving the hypoxia problem.

I see Don Boesch's point and agree with him that the Bianchi *Eos* articles went too far in criticizing the focus on limiting nutrient inputs as the way forward to solving the dead-zone problem. I think we're right to focus on nutrients even given the severity of wetland loss in Louisiana. But it's too much to expect that oceanographers and other scientists will stop exploring aspects of the problem that are not part of the standard story about why dead zones form. Scientists are always tyrants who want more data and greater understanding. Unanimity is rare. This reminds me of the climate change debate in the United States. In the 1980s, some politicians did use disagreements among scientists to argue against doing anything about the problem, pointing to the need for more research. Although the argument was wrong, I don't think it was irrational. But now the research has been done, and the results are in; there is a solid consensus among scientists about the severity of climate change and the urgency to act. Even with the new understanding and near-unanimity among scientists, however, too many Americans think climate change is not that serious or perhaps is even a hoax. Public opinion about these issues is shaped by more than just science and the "facts."

Nutrients and Pollen in the Chesapeake

The dead zone origin story for the Chesapeake Bay is similar to what we just saw for the Gulf of Mexico. Data go back further for the Chesapeake and are more extensive, but the message is about the same, starting with the role of freshwater input. The flow of the Susquehanna River, the largest feeding into the Chesapeake, has gone up and down over the years, as has the hypoxic area. As discussed in Chapter 2, Seliger and colleagues were right about high river flow and the lack of strong winds causing the catastrophic anoxia in 1984. But that's just one year. There is no overall upward trend in river flow from 1884 to 2012 at least.[28] Precipitation in the Chesapeake watershed has increased over this period, but it hasn't resulted in more fresh water in the Bay due to the weak relationship between rainfall and discharge into the Bay.[29] As with the Mississippi, the flow of fresh water into the Chesapeake may increase in the upcoming decades,[30] but that's in the future. The spread of hypoxia in the Chesapeake over the last fifty years is not because of changes in freshwater flow.

What changed were the nutrients, both in the Bay and its tributaries. As with the Mississippi, in the Susquehanna River, concentrations of the important nutrient nitrate have tripled from 1950 to now.[31] With more nutrients coming in from the Susquehanna, more has been seen in the Chesapeake itself; nitrate has doubled. With more and more nutrients, there has been more and more algal production and less and less oxygen dissolved in Chesapeake bottom waters. As in the Gulf of Mexico, nutrients explain the expansion of the Chesapeake dead zone in the 20th century.

But we left dirty footprints before then. As with the Gulf, we can use sediment time capsules to go back further in time and see those footprints in the Bay.[32] The Chesapeake began to be transformed by Europeans settling on its shores and in its watershed, much earlier than seen with the Gulf. Maryland, Pennsylvania, and the rest of today's states in the Bay's watershed were among the original thirteen British colonies established in the 17th and 18th centuries. Early Europeans encountered a land covered in forests and wetlands, recorded in the pollen from trees and water lilies preserved in Chesapeake sediments. The tough walls of pollen, the allergy-causing spores of seed plants, differ in shape enough for ecologists exploring ancient life (paleoecologists) to distinguish among the types of vegetation that flourished hundreds or thousands of years ago. Even the pollen from closely

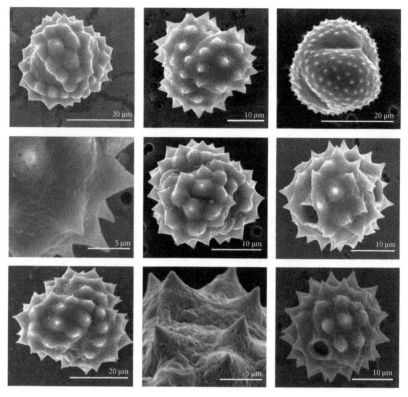

Figure 4.2 Examples of pollen. Pollen from other plant species have been used to deduce vegetation in the past. Provided by Joan Vallès and used with permission of the publisher.

related species varies greatly, as illustrated in Figure 4.2.[33] Pollen preserved in Bay sediments show a shift from oak to ragweed as forests were cut down for pasture, starting in 1760, about thirty years after the founding of Baltimore, the largest city at the time in the Chesapeake watershed.

By the mid-1800s, as settlers continued to cut down forests and to turn grasslands into croplands, erosion increased, leaving another mark on the Bay still evident today. Chesapeake waters likely became cloudier starting in the 19th century, according to diatom silica shells preserved in sediments. One type of diatoms, the cigar-shaped pennates, live on the bottom in shallow waters and were most abundant in pre-European sediments, and then another type, centric diatoms, which look like circular pillboxes, took over in the mid-1800s. One explanation is that centric diatoms, which live suspended in the water column, increased in abundance as nutrient inputs increased, while bottom-dwelling pennates died off as they became shaded by cloudy, more turbid waters above them. The turbidity was caused by soil eroding from land and by lush algal growth in surface waters. Turbidity also helps to explain the death of seagrasses in the Chesapeake. These large vascular plants are normally rooted in shallow sediments of the Bay and depend on light penetrating through the water above. In addition to producing oxygen like all plants, seagrasses provide food and shelter for many fish, shellfish, and turtles. Like pennate diatoms, seagrasses die in gloomy, turbid waters. Paleoecologists have used seeds preserved in sediments to follow different species of these submerged plants over hundreds of years.[34] The full story is complicated, but the brief version is that seagrasses were abundant before European colonization, less so after. These plants really declined around 1970.

The sediment record indicates that eutrophication of the Chesapeake started in the 1800s and intensified in the 20th century. At the same time, pollen and seeds and other data record the transition from an estuary with abundant submerged plants and benthic organisms to one full of algae, fueled by nutrients from land. The rest of the story should now be familiar. The mid-century increase in nutrients led to more algae and more organic material and less oxygen. The end result is the dead zone we see today in the Chesapeake.

Nutrients and Major Baltic Inflows

Hypoxia was soon linked to nutrients in many dead zones around the world, such as the Adriatic Sea,[35] the Black Sea,[36] and the East China Sea.[37] Even earlier than in the Gulf of Mexico, Swedish scientists made the link between nutrients and oxygen levels in the Baltic Sea. In the late 1960s, S. H. Fonselius thought that the low oxygen content of Baltic bottom waters was due to increasing concentrations of one nutrient, phosphate.[38] More recent research established that phosphate increased by eight times from 1900 to its peak in the mid-1980s, while nitrate increased by four times over the same period.[39] As the nutrients increased, the dead zone in the Baltic metastasized to the huge environmental problem of today.

What about changes in the physics? Some physical oceanographers argued that the size of the Baltic dead zone was controlled by delivery of dissolved oxygen from the North Sea. Perhaps that delivery has changed over the decades. To explore that possibility, we need to review how oxygen gets into Baltic bottom waters.

The Baltic is quite different from the Gulf of Mexico and the Chesapeake Bay. In the two North American dead zones, bottom waters eventually get dissolved oxygen in autumn when stratification is broken down as air and surface water temperatures decline and winds pick up, forcing the mixing of the top and bottom layers of the water column. That brings oxygen to bottom waters and ends the hypoxia. Only some of that autumn mixing occurs in the Baltic. The upper layer of the Baltic is mixed in autumn when temperatures cool, but the bottom layer remains isolated from the top by a gradient in salt, a permanent halocline. That means hypoxia in the bottom layer is also permanent, with the halocline cutting off the vertical supply of dissolved oxygen from the surface. Diffusion is too slow to help. However, occasionally bottom waters of the Baltic get some oxygen from an influx of oxygen-rich, salty water from the nearby North Sea and the more distant North Atlantic Ocean. The North Sea water is dense and hugs the bottom as it enters the Baltic Sea. This influx is limited, however, because of the tortuous connection between the Baltic Sea and the North Sea (Fig. 4.3), consisting of three narrow straits that thread their way past Danish islands or between Denmark and Sweden, separated by only a kilometer at some points. Flow via the Danish straits is restricted even more by shallow sills sticking up from the bottom. Salty, oxygen-rich oceanic water only occasionally enters into the Baltic in any appreciable amount. When it does, it's a major event, so much so it gets a special name: Major Baltic Inflows (MBIs).

At first it was thought that MBIs would alleviate hypoxia in Baltic bottom waters, and there was some evidence of that happening. But then Daniel Conley, now a professor at Lund University in Sweden, gathered more data and came to a radically different conclusion.

I recently met up with Daniel near my hotel in Lund and walked with him past old university buildings to his office where we talked about his work on the hypoxia problem. Although he could pass as Nordic, with his thinning blond hair, Daniel describes himself as a "mutt of an American."[40] He grew up in Ft. Lauderdale, Florida, and went to college not far from the Gulf's dead zone, at Tulane University in New Orleans. He did a Master of Science degree at the University of Wisconsin-Green Bay, once known as "Eco-U" for its unique focus on the environment, where he worked on the low oxygen problem in Green Bay (the body of water, not the city). But his real education

Figure 4.3 Baltic Sea, with its tortuous connection to the oceans. The input of new oceanic water from the North Sea is called a Major Baltic Inflow (MBI).

and training in the field came during his PhD work at the University of Michigan, where he studied silicate, that crucial nutrient for a crucial type of algae, diatoms. Daniel then went on to the Horn Point Lab, part of the University of Maryland system, where he worked on eutrophication, the input of excessive nutrients.

Conley and colleagues looked at the relationship between one plant nutrient, phosphate, and the size of the Baltic dead zone; but it was their ideas about MBIs that attracted fierce opposition.[41] Previous work suggested that the paucity of MBIs since the mid-1970s had led to stagnation in the bottom layer and severe hypoxia. Conley, however, found that the size of the Baltic dead zone was smallest at the end of the stagnation period, which is exactly opposite from what you'd expect if MBIs have such a big impact. Daniel mentioned to me that the paper and he got a lot of flak, especially from oceanographers at a certain German institute (he didn't divulge names). His 2002 paper went through two rounds of review before being rejected by the prestigious journal *Nature*, although it was eventually published in another excellent, albeit less prestigious, outlet, *Environmental Science and Technology*. However, over time, others came around to Conley's view. Everyone was especially convinced by the Christmas MBI of 2014.

In December of that year, a strong MBI surged into the Baltic, the third strongest on record since 1880.[42] It was the first to occur in the previous 10 years, the last two being in 1993 and 2003. The initial report predicted this unusually strong (and rare) MBI would "most probably turn the entire Baltic deep water from anoxic to oxic conditions." And the new water did initially bring some relief to the Baltic dead zone, but it was short-lived. Already by the fall of 2015, not even a year after the Christmas MBI, oxygen was again gone,[43] and the Baltic was far from being completely oxygenated as initially hoped. In fact, the Christmas MBI appeared to have stimulated oxygen use in bottom waters because the new water brought along more organic material for bacteria to degrade.

Jacob Carstensen, who led a seminal study about the spread of the Baltic dead zone discussed in Chapter 3, pointed out to me that an MBI can push hypoxic waters into areas of the Baltic that previously had plenty of oxygen. An MBI also brings salt with it, which strengthens stratification and further reduces mixing of hypoxic bottom waters with the oxygen-rich surface. So, Conley was proved right. MBIs do bring in dissolved oxygen, but they also bring in a lot of trouble for the Baltic. Jacob thinks that if MBIs were more common, they would do more good by continuously circulating water like the water movement created by a river and tidal action in an estuary. But now the Baltic Sea is more like a lake, interrupted too infrequently by MBIs that do more harm than good.

Regardless of whether MBIs have a positive or negative impact on oxygen in the Baltic, the important question is whether they have become more or less common over the last 100 years. If they have, then they are candidates for explaining why the dead zone expanded in the Baltic in the 20th century. Maybe nutrients aren't totally to blame. It is true that there have been few MBIs in the last twenty years, but that's too late. The Baltic dead zone was already permanent and extensive long before 2000. Looking back further in time, to 1880 when data for the Baltic were first collected, it appears that MBIs have come and gone over the years; every 20 to 30 years there have been more MBIs and every other 20 to 30 years there have been fewer for reasons that aren't entirely clear.[44] But overall, there is no trend up or down in MBI frequency. That brings us back to the definite trend up in nutrient levels as the best explanation for the expansion of the Baltic dead zone in the 20th century.

The lack of a trend in MBIs also indicates that changes in climate since 1880 can't explain the Baltic dead zone. But climate effects can be seen if we go back further in time, by looking for clues left in Baltic sediments.

The clue used by scientists this time was lamination,[45] the orderly layering of material that has settled to the seafloor. Lamination can only occur in the absence of currents and of large benthic animals, both of which stir up sediments and disrupt the layers. No currents mean no oxygen and no oxygen-loving bioturbating animals, allowing sediments to go undisturbed and lamination to form. In 1997, Swedish scientists on the *R/V Kotsov* took a nine-meter-long sediment core from the Gotland Basin, one of the main lobes of the Baltic Sea. The core was long enough for the scientists to go back about 10,000 years before the present. The researchers found lamination on and off through the sediment core, indicating hypoxia had come and gone through the millennia. These hypoxia episodes result from natural causes intertwined with human-forced changes on Baltic shores.

The earliest sign of hypoxia appears in sediment layers deposited between 8000 and 4000 years or so ago, with a brief (for geology) respite around 6500 years ago.[46] During this period of hypoxia, local temperatures were relatively high, deduced by scientists based on pollen in lake sediments and other data, indicating that glaciers had receded from Scandinavian mountains and the pine-tree line had climbed up as much as 400 meters higher than today. Temperature itself is not thought to be the prime mover of hypoxia. Rather, mild temperatures may have directly or indirectly (via precipitation) contributed to high stratification that cut off oxygen's flow to bottom waters of the early Baltic. So, climate change contributed to the creation of this prehistoric dead zone. But changes in the geology around the Baltic probably had an even larger effect on stratification and oxygen levels.

This early period of hypoxia occurred when the Baltic Sea was transitioning from a freshwater lake, Ancylus Lake, to the Littornia Sea, caused by the inflow of salt water via the Öresund Strait, starting about 8000 years ago (the exact date is debated) when the planet was still recovering from its last ice age. About 10,000 years ago, the level of the oceans rose due to melting ice and warming temperatures, and the land sprung back up after being weighed down by glaciers. Ancylus Lake was transformed into a sea by the inflow of salt water from a rising North Sea and the uneven rebounding of land surrounding the nascent Baltic. As a result, the salty water coming through the Öresund Strait led to strong stratification and the depletion of oxygen in bottom waters. The brief respite of oxic waters between 7000 and 6000 years ago coincided with a small drop in temperature, more precipitation, and less salty conditions. So, climate change explains the short period of oxygen-rich waters, and changes in both climate and geology explain the persistent, early hypoxia in the Baltic starting 8000 years ago.

Climate change was a major force behind the next swing in Baltic's oxygen levels. Hypoxia's hold on the Baltic was broken about 4000 years ago when temperatures again began to fall. The tree line came down, and glaciers returned to Scandinavian mountains. The changes in temperature, precipitation, and other climate properties that vary along with temperature and precipitation led to lower biological production and less organic material that could deplete oxygen. Climate change affected stratification but so too did geology. Around 4000 years ago, as the land continued to rebound, the connection between the North Sea and the Baltic tightened, slowing the influx of salt water and weakening stratification. Sediments were jumbled, without lamination for about 2000 years, as oxygen-rich waters brought back benthic fauna and bioturbation.

And then temperatures swung the other way, rising again, starting about AD 950 and lasting to the beginning of the 13th century. During this Medieval Warm Period, the Norse grew wheat on southwestern Greenland and lamination returned to Baltic sediments, indicating the loss of oxygen from bottom waters. One recent study argued that this hypoxic period was caused by climate change,[47] but Lovisa Zillén and Daniel Conley argue that people share at least some of the blame. They point out that the human population in the Baltic watershed nearly doubled during this period, to roughly 9 million in 1300 from 4.6 million in the year 1000. The burgeoning population was supported by more intensive agriculture, as forests were cut and grasslands cleared to make way for crops. The Roman plow was replaced by the soil-inverting moldboard plow, exposing more soil to erosion by wind and rain, releasing still more nutrients that eventually flowed into the Baltic. There, nutrients stimulated algal growth and the production of organic material that drives oxygen consumption. The result was a medieval dead zone.

The return of oxygen around 1300 was good news for the Baltic Sea but was the result of bad news, to say the least, for people living in the Baltic watershed and the rest of Europe. The year marked the start of the Little Ice Age when summers were cool and winters colder. Glaciers advanced down mountain valleys of the Alps and Norway, destroying farmhouses and small villages and smothering alpine meadows where livestock once grazed.[48] Norse settlements in Greenland were abandoned. The growing season was short, and harvests failed. As if that wasn't enough, Black Death then swept in from Asia, killing about half of Europe's population in the 14th century. Sweden's population plummeted by nearly 70 percent, to 347,000 in 1413 from over a million at the end of the Medieval Warm Period in 1300. As the Little Ice Age continued, and famine and disease marched on, farms were deserted, and fields went fallow. This devastation of agriculture meant fewer plant nutrients leached into the Baltic Sea. Fewer nutrients meant lower organic material production and more oxygen in Baltic bottom waters.

The laminated sediments tell one more story about oxygen in the Baltic, starting about 1900. The Little Ice Age has ended, but another age, the Industrial Revolution, is just getting started. The number of people living in the Baltic watershed has grown exponentially since the dark days of the Black Death, and agriculture has expanded and intensified even more so. The result is the return of lamination to Baltic sediments and of hypoxia to Baltic bottom waters. Now the laminated sediments story overlaps with direct data of dissolved oxygen in the Baltic, clearly showing the spread of the dead zone in the 20th century.

The sediment records indicating past hypoxia remind me of the current debate about the start of the Anthropocene. The term has been proposed by some geologists to describe the epoch when humans have impacted the Earth on a grand scale, on par with glaciers and asteroid strikes. One suggested start date for the Anthropocene is very precise: 5:26 in the morning of July 16, 1945. That's when a nuclear weapon was first tested in New Mexico. The year is also about when dead zones started to proliferate around the world. There are the earlier signs of human impacts, preserved in sediments, of changes in terrestrial and aquatic flora, water quality, and dissolved oxygen levels. Many of these signs are subtle, recognizable only by the expert, certainly less dramatic than the radioactive remains of a nuclear bomb. Regardless, they are evidence that estuaries, seas, and coastal waters were sullied by human activity hundreds of years ago when the technology was primitive and the land sparsely settled. But hypoxia was spotty, not persistent or widespread, before the mid-20th century. It was then that nutrient levels started to creep up, and dead zones began to expand and proliferate.

5

Giving the Land a Kick

Something happened midway through the 1900s that eventually led to the hypoxia problems of today. Given everything going on during those tumultuous years, scientists and everyone else can be excused for not noticing the signs. There were only hints in scattered technical reports and papers buried in the scientific literature. Now we can piece together the evidence and see a change midway through the century; nutrients started to increase in the streams and rivers that eventually empty into today's dead zones, like the northern Gulf of Mexico, the Chesapeake Bay, and the Baltic Sea. Roughly at the same time, bottom waters of these waterbodies started to run out of dissolved oxygen. It's pretty clear that we did something midway through the 1900s that elevated nutrient levels. What happened? Where did those nutrients come from?

We need to revisit the sewage question. In addition to the sewage organics that directly fuel oxygen use by bacteria, sewage contains nutrients like ammonium and phosphate that would indirectly cause oxygen depletion. The nutrients are taken up by algae, which make organic material that is readily used by oxygen-consuming bacteria. These nutrients are released during the decomposition of the sewage, either in the treatment plant or in the waterbody receiving the plant effluent; except for algal uptake, the nutrients are removed only by tertiary treatment of the wastewater, not by the more common and less expensive primary and secondary treatments. Even when the sewage organics are not a problem, sewage can be an important source of nutrients, packing more of a punch than its organic material. One molecule of sewage organic carbon uses up just one molecule of dissolved oxygen, while one molecule of a sewage-derived nutrient could lead to the production of seven to a hundred molecules of organic carbon, which then go on to deplete seven to a hundred molecules of dissolved oxygen.[1]

So, a little bit of sewage goes a long way in stimulating algal growth and eventually using up dissolved oxygen. But sewage isn't a big source of nutrients for most of the dead zones we've discussed so far. In rich countries, these nutrients are removed by wastewater treatment, aided by regulations banning the use of phosphate in detergents. Municipal sewage, runoff from cities, and industrial sources today contribute 5 to15 percent of total nitrogen that enters the Gulf of Mexico,[2] Baltic Sea,[3] the Black Sea,[4] and the Chesapeake Bay.[5] Likewise, these sources contribute about 15 percent of the phosphorus going into the Gulf and the Chesapeake. The numbers are a bit higher for the Baltic Sea and the Black Sea (24 and 29 percent, respectively), but sewage generally isn't the main source for these dead zones.

If not sewage, where are the nutrients coming from? Why did levels of nutrients increase so much midway through the 20th century? For the dead zone regions just mentioned, the answer is agriculture. Let's first focus on the Gulf where the case against agriculture is the clearest.

Corn Feeds a Dead Zone

The story for the Gulf dead zone origin needs to begin with a map of the Mississippi River and the land it drains. The watershed of the Mississippi is huge, including all or parts of 32 US states and even bits of two Canadian provinces, forming a large inverted triangle that funnels down to the mouths of the Mississippi and Atchafalaya Rivers debouching into the northern Gulf of Mexico. Now superimpose onto the basin map the Corn Belt states (Fig. 5.1) that produce nearly all of the corn (maize) in the United States. The country as a whole accounts for about a third of all the corn produced in the world, worth over $50 billion per year.[6] Five Corn Belt states (Illinois, Indiana, Iowa, Minnesota, and Ohio) alone grow about half of all corn in the country. One may wonder why the town of Olivia, Minnesota, is home to the Corn Capital of the World when Iowa leads the nation in corn production. But if the world is to have a corn capital, no doubt it belongs in the Mississippi River watershed.

There is much more to the Corn Belt than corn. The Corn Belt is also the Soybean Belt, in a country that produces more of the grain than any other does. Those five Corn Belt states account for almost as much of soybean production as they do corn: 48 percent of the US total in 2017. An even higher fraction, 70 percent, of soybean production is in the thirteen states of the Mississippi basin critical conservation area; in addition to the five Corn Belt states, the critical area includes Arkansas, Kentucky, Louisiana, Mississippi, Missouri, Tennessee, parts of Wisconsin, and the eastern edge of South Dakota.[7] Along with growing a lot of corn, Iowa leads the United

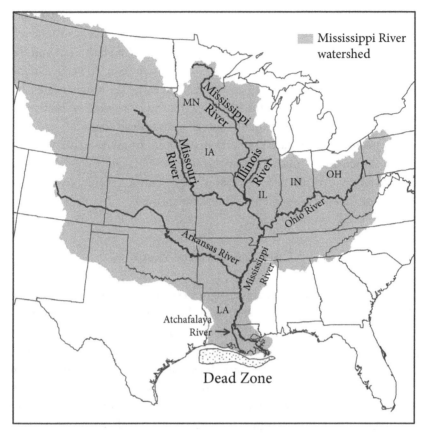

Figure 5.1 The Mississippi River, some of its major tributaries, and its watershed, including five Corn Belt states: Iowa (IA), Illinois (IL), Indiana (IN), Minnesota (MN), and Ohio (OH). The Gulf of Mexico dead zone is mainly in coastal waters of Louisiana (LA).

States in soybean production—and also in raising hogs and pigs. In spite of that, Illinois claims world capitals for both commodities: soybeans in Decatur,[8] porcine animals in Kewanee. Even if they don't have any world capitals—except for Belzoni, Mississippi, Catfish Capital of the World—the states along the Mississippi River south of the Corn Belt are no agricultural slouches. Arkansas, for example, produces more rice than any other state and ranks third in poultry production.[9] Many amber waves of grain are grown, and cattle, hogs, and chickens are raised in the Mississippi River drainage basin.

There is a reason why so much corn and everything else is in the Mississippi basin. The river and its tributaries are responsible for the region's rich soil. The raw material for that soil came from rocks that had been abraded by the Laurentide Ice Sheet, which covered most of Canada and the northern United States between 95,000 and 20,000 years ago.[10] When the ice sheet finally melted, fine-grain sediment was carried south by the Mississippi River, then a massive braided river with tentacles extending throughout the center of North America, from the current border between Canada and the United States down to the Gulf. Sediments were deposited along its banks and flood plains, forming nutrient-rich bottomlands and intervales. Glacial sediments were also carried in by strong west and northwest winds to form the loess hills of western Iowa and loess bluffs on the eastern side of the Mississippi. The result is some of the most productive soils on Earth.

Those rich soils may be the source for nutrients ending in the Gulf. The great naturalist Aldo Leopold observed, "All land represents a downhill flow of nutrients from the hills to the sea."[11] We have seen that human activity started to hasten that downhill flow to the Gulf of Mexico by the 19th century. The early eutrophication of the Gulf and other water bodies was caused by soil erosion and leaching of nutrients when forests and prairies were transformed into cropland and pastures. Nutrients from eroding soils undoubtedly damaged some waterbodies, fueling the sporadic hypoxia that occurred before the mid-1900s. Perhaps soil erosion and leaching somehow increased greatly after then and caused nutrients in dead-zone tributaries to shoot up around 1950.

But the timing just isn't right for this to be the explanation. Moldboard plows replaced hoes and ards long before low-oxygen waters became commonplace, and the transformation of forests and grasslands to farmlands was largely complete before the rise of the dead zone in the Gulf of Mexico. From 1860 to about 1950, the area covered by crops in the United States went up over nine times, but after 1950, it declined by about 15 percent.[12] For many states along the Mississippi River and in its basin, the increase was more dramatic and the decline less severe (or the area increased somewhat), but the implication is the same. From 1860 to 1950, the acreage devoted to farms doubled in Illinois and expanded nearly ten times in Iowa; but that acreage increased by only 40 and 26 percent, respectively, from 1950 to 2012.[13] Connected to the Mississippi River via the Arkansas River, Kansas didn't enter the Union until 1861 and then only after violent fights between those who were pro-slavery and the abolitionists in the years leading up to the Civil

War. In spite of its contentious pre-Union status, the US census still surveyed Kansas in 1860 and found 405,468 acres under cultivation. (New York had 14,358,403 acres of farmland, the most in the nation in 1860.) Kansas crop-land increased by about 70 times and reached its zenith in the early 1930s, after which it decreased by about 10 percent.[14] For the entire critical conservation area of the Mississippi River, the rapid increase in crop coverage was before 1920, after which farmers added only about 10 percent more cropland, just as the Gulf of Mexico dead zone was expanding.

So, soil erosion and leaching alone don't explain the rise in nutrients and the proliferation of low oxygen waters in the Gulf around 1950.[15] More precisely, the nutrients that had been in the soils for eons or added there by 19th-century farming practices were not the nutrients causing eutrophication in the northern Gulf of Mexico during the second half of the 20th century. It doesn't make sense that the input of these old nutrients somehow increased around 1950 when their source, farmland, only marginally grew in area. Something else must have happened in the Mississippi River basin in the middle of the 20th century.

What happened was farmers got really good at growing corn and everything else. Although agricultural production increased from the end of the 19th century to the beginning of the 20th century, it really shot up around 1950. The first increase in corn production, from the 19th century to the first decades of the 20th century, was made possible by the conversion of forests and grassland into cornfields, not because farmers got better at growing the grain (Fig. 5.2). Roughly the same number of bushels were harvested from each acre in the mid-19th century as in the early 20th century. The yield was 24.3 bushels per acre in 1866, the first year when data were collected, and 28.9 bushels in 1940. But then, corn yields started to improve and production skyrocketed. By 1950, the average yield was 38 bushels per acre, and then it zoomed to over 170 bushels today. This timing is strong circumstantial evidence damning agriculture. During the same decades that the farmers got better and better at growing corn (and everything else), nutrient concentrations got higher and higher in the Mississippi River and its tributaries. Those same years saw the proliferation of the Gulf dead zone.

Several changes in agriculture contributed to the higher yields. The adaptation of hybrid corn, developed by old-fashioned plant breeding, led to the first upswing in yield starting in the late 1930s.[16] Herbicides and pesticides were then introduced to control weeds and insect pests, and farmers switched from horses to tractors. More corn plants of the new breeds could

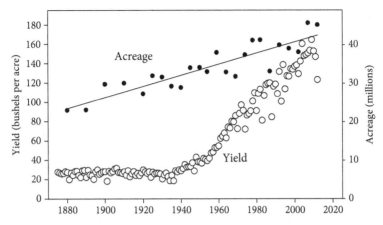

Figure 5.2 The land devoted to corn in five Corn Belt states (Illinois, Indiana, Iowa, Minnesota, and Ohio) doubled from 1880 to 2012, while corn production increased about fourfold because yields improved, starting around 1940. The acreage data are from the USDA Census of Agriculture. The yield data are from the USDA National Agricultural Statistics Service (ISSN: 2157-8990).

be planted on an acre, closer together and in tighter rows. The herbicides and pesticides, further mechanization, and more improvements in corn genetics made possible a second, much larger increase in corn yields starting around 1955. Contributing to these upswings in yields was another group of chemicals, fertilizers.

"Whip to the Tired Horse"

The Corn Belt and much of the Mississippi River basin was the frontier back in the 19th century. It wasn't yet the Midwest—it was just the West. Early settlers sent back East tales of a "land that would never wear out," where all you had to do was drop a seed into the rich loam in spring and harvest gold in autumn. "There grass grows as luxuriant as hops, so that in a single day you can gather winter fodder for your cow, big as an elephant . . . [you] never need manure the field," according to a Norwegian emigrant in 1844.[17] Except perhaps for elephant-size cows, reports of the fertile land were largely true at first. Although eventually fields needed manure and crops rotated to maintain yield, only farmers who did not follow these practices had to resort to commercial fertilizers, or so it was thought by some.[18] Commercial fertilizers were like a "whip to the tired horse," used only by careless farmers. In the late 19th century, commercial fertilizers were mostly organic, from the remains of meatpacking or rendering factories: blood, bones, and offal from slaughtered cattle and hogs.[19] Guano from Chile and peat were added to the mix. Because of their cost, fertilizers were used only for a few, highly profitable truck crops like tobacco, tomatoes, lettuce, and melons, not on corn or other row crops that covered most of the land. By the 1930s, although agriculture was beginning to change, fertilizer use was still low. It wasn't cost-effective when crop prices were so low during the Great Depression. All changed with World War II.

Starting in 1940, fertilizer use soared. By then, farmers realized that fertilizer "gives the land a kick"[20] and could boost yields to the high levels needed to feed the war effort. For the entire country, the use of phosphorus fertilizer increased by six times and nitrogen fertilizer by nearly thirty times from 1940 to 1979,[21] the year when fertilizer use started to level off. Application of potash, another important plant nutrient, increased by a factor of thirty during the same time. The transformation was even greater in the Mississippi watershed. The highest rates of fertilizer use shifted from the southeastern and eastern United States to the Great Plains, the Northwest, and the Midwest—parts of the country with the closest ties to the Mississippi River.[22] What area of the United States used the most?: the Corn Belt. Five Corn Belt states—Illinois, Indiana, Iowa, Minnesota, and Ohio—accounted for only about 4 percent of total nitrogen fertilizer use in the country in 1940 (378,543 tons) but over 30 percent in 1979 (10,549,900 tons). The numbers are even higher for twelve critical conservation area states in the Mississippi River Basin. (I excluded the sliver of South Dakota for this analysis.) These states accounted for 15 percent of the nitrogen fertilizer used by the country in 1930 and 44 percent in 1979. Use of that fertilizer in the critical states went up nearly 60 times, to 4,639,400 tons in the late 1970s from 53,626 tons in 1930. In 2011, the total was 5,715,120 tons.[23] As much as half of the increase in corn yields in the mid-20th century has been attributed to fertilizers.[24]

All that fertilizer applied to the fields did not, and does not, stay there. Figuring out how much leaves is complicated. First off, depending on the type and variety, crop plants only take up a fraction of the applied fertilizer. The estimates vary greatly, but it's roughly 50 percent,[25] although one study put the phosphorus use efficiency as low as 16 percent.[26] What happens next to the unused fertilizer depends on many factors.[27] Some of it stays stored in soils at least temporarily, while some leaches out, depending on the fertilizer element (nitrogen loss is not the same as phosphorus loss) and chemical form (the fate of anhydrous ammonia is not the same as the fate of ammonium nitrate). Farmers have more time to apply fertilizer in late fall or winter, but loss is greater when fields are bare or fallow. Broadcasting fertilizer onto a field is easier and cheaper, yet less is lost when the fertilizer is injected below the soil surface by coulter or air. Not tilling a field, the definition of "no-till agriculture," may cut down on soil erosion, but it leaves fertilizer more exposed and subject to greater loss. Fertilizer loss is negligible from rice paddies but extreme for farms on steep tropical slopes. Not much is lost when it doesn't rain, much can be when it pours.

Even a small fraction of fertilizer leaching from a field can have a big impact on rivers, lakes, and eventually coastal oceans. If a cornfield in Iowa loses only 0.5 percent of the 40 to 64 kilograms of phosphorus fertilizer applied per hectare (36–58 pounds per acre), water from the field could contain 200 to 500 micrograms of phosphorus per liter.[28] There are reports of concentrations reaching 20,000 micrograms.[29] That phosphorus-rich water potentially flows into ponds, reservoirs, lakes, and rivers with much lower nutrient levels. The cornfield water is ten to a hundred times more nutrient-rich than a pristine stream or the surface waters of the northern Gulf of Mexico. And because that cornfield oozes nutrients on and off all year, its impact is greater than implied by the difference in concentrations. So, even a small fraction of fertilizer kicking the land can end up giving a big kick to nearby water bodies.

One of the first signs that fertilizer was a problem was the correspondence between fertilizer use on land and nutrient levels in dead-zone tributaries.[30] There is an incredibly strong relationship between nitrate in the Mississippi River[31] and use of nitrogen fertilizer[32] between 1960 and 1988 (Fig. 5.3). There is also an incredibly strong relationship between the nitrogen from fertilizer and other anthropogenic sources and the nitrogen flowing down rivers in different parts of the world.[33] Northern Canada, far from farms and big cities, receives little nitrogen from fertilizer and other human sources, and its rivers are pristine, with low nitrogen fluxes. At the other extreme, the highest input of human-made nitrogen is in the watersheds surrounding the North Sea, which is connected to the Baltic Sea. Although sewage is a big contributor of nutrients thanks to the large number of inhabitants living in the North Sea watershed, agriculture, and particularly fertilizers, are the biggest source. The correlations indicate that more fertilizer leads to more

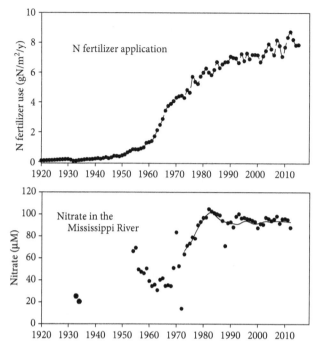

Figure 5.3 Nitrogen (N) fertilizer application (grams of nitrogen applied per square meter per year) in ten critical conservation area states in the Mississippi River watershed (top panel) and nitrate concentrations in the lower Mississippi River (bottom panel). The fertilizer data are from the US Department of Agriculture, and the nitrate data are from Gene Turner and the US Geological Survey.

nutrients in rivers, including the Mississippi flowing into the Gulf of Mexico dead zone.

Most studies have focused on nutrients from fertilizers, but another agricultural source is manure. The connection between manure and nutrients has been explored in the Baltic Sea watershed,[34] the Chesapeake Bay watershed,[35] and Iowa,[36] a state that sends a lot of nutrients to the Gulf of Mexico. In areas of these regions with intense animal husbandry, nutrient levels in streams and rivers are higher than elsewhere, implying manure is the cause. Chris Jones led the study about manure in Iowa and in a blog has looked at manure production from a different perspective.[37] He calculated Iowa's "real population" in terms of waste excreted by livestock as well as people. Being a fairly big mammal, we produce a lot of waste, but nothing like a hog. Even though it is about the same size as an average Iowan, each hog excretes three times more nitrogen and five times more phosphorus. Manure production by other livestock isn't quite as impressive, but they do their part. So, the real population of Iowa is more than just the three million humans living in the state; in terms of waste production, we also have to count the state's 20 to 24 million hogs, 1.8 million beef cattle, 250,000 dairy cattle, 80 million laying chickens, and 4.7 million turkeys. Jones put the numbers together and calculated that Iowa's real population is equivalent to 134 million people, which would make it the tenth most populous country in the world, just below Russia. One watershed alone, the Raccoon North, has the equivalent of over nine million people, about the same as Tokyo.

Manure doesn't change the argument that fertilizers are the source of nutrients ending up in dead zones like the Gulf of Mexico. Nitrogen and phosphorus in manure start as nitrogen and phosphorus in fertilizer, which is used to grow the grains fed to those millions of hogs, cattle, and other livestock. Iowa leads the United States in raising hogs because it is so good at growing corn. Putting hogs close to their feed cuts down costs. As Chris Jones says in his scientific report, "Co-locating crop production and livestock within the state has created efficiencies of production, transportation, and fertilization."[38] He thinks manure is a bigger source of nutrients contaminating Iowa waterbodies and eventually the Gulf of Mexico than now suspected. Even if fertilizer is still the main direct source of those nutrients, we need to remember manure in thinking about how to reduce nutrients from agriculture.

Big Agriculture Fights Back

The accusation that agriculture in general and fertilizers in particular caused the Gulf dead zone created a stir in the 1990s. In retrospect, the link between fertilizers and dead zones shouldn't have been a surprise. Scientists working in lakes pointed out the connection between agriculture and eutrophication in the 1960s, and soon after, as sewage treatment improved, it became apparent that nutrients from agriculture were causing eutrophication of the Baltic Sea.[39] It's not hard to see that if you use a lot of fertilizer, some of its going to end up in nearby streams and rivers. Nutrients are bound to leach away from mounds of manure. So, it should not have been controversial to discover that agriculture was a big source of nutrients fueling hypoxia in the northern Gulf of Mexico.

And perhaps the controversy would have been limited to muted, double-hedged exchanges among scientists if not for the work of Nancy Rabalais and Gene Turner, that wife-husband team working on the Gulf's hypoxia problem. Their first salvo in the controversy was the study mentioned before, the one published in 1991 pointing out the correlation between fertilizer use and nutrient levels of the Mississippi River. Then in 1993, using data from the just-completed summer cruise to survey Gulf oxygen concentrations, Rabalais put out a press release to announce the size of the Gulf dead zone that year. She wanted to get the word out because the dead-zone area in 1993 was twice as large as she had seen since the survey cruises started in 1985. Her sounds of alarm led to scientific meetings and task force formations and eventually the establishment of a White House advisory body, the Committee on the Environment and Natural Resources (CENR).[40] One of its first jobs was to review the hypoxia problem and make some recommendations.

The CENR assessment outlined the evidence we've already seen linking fertilizers with nutrients in the Mississippi and hypoxia in the northern Gulf of Mexico. (Most of the material had already been published in peer-reviewed journals. Scientists generally published their best stuff in these journals and then repurpose it for government reports. Unlike many reports, however, the CENR publications were peer-reviewed—vetted by independent experts in the field.) Appropriately, the first of the six CENR reports was written by Rabalais and Turner, along with three other coauthors.[41] Rabalais and coauthors discussed the changes in plant nutrients, especially the increase in nitrate in the Mississippi River from the mid-1950s to the late 1980s. They reviewed the changes in land development during the 19th century, then segued into the biggest change that affected nutrient fluxes: the rise in fertilizer use in the 20th century. Another CENR report was more specific about where those fertilizer nutrients are coming from.[42]

The US states thought to be the biggest sources should now be familiar: the five Corn Belt states already highlighted for the bushels of corn and soybeans they grow, the number of hogs and pigs they raise, and the kilograms of fertilizer they use to achieve that high productivity. Those five states accounted for 56 percent of the nitrate going into the Mississippi River according to the 1999 CENR report. Of the five, Iowa and Illinois stood out. Although making up only 4.5 percent of the Mississippi River basin area, the two states contributed about 35 percent of the nitrate released into the river. Some of that nitrate may not make it all the way to the Gulf, but much of it does because it is not used by algae in the river. The CENR report suggests that, in order to shrink the dead zone, farmers need to adapt "best management practices" to reduce fertilizer use and loss.

The advice wasn't well received in the Corn Belt. Farmers and the fertilizer industry saw in the CENR assessment the threat of new regulations clamping down on fertilizer use. In a speech at the annual meeting of the American Farm Bureau Federation in January 2000, the chief of the Illinois State Water Survey at the time, Derek Winstanley, was quoted as saying the report authors were steeped in "environmental religion" and new regulations were "a big steam train coming down the track . . . right at you."[43] Later that year, he and Edward Krug put out a 172-page report (not counting a 15-page preamble and another 12 pages of additional notes), arguing that "the Illinois River system was hyper-trophic [very nutrient-rich] prior to the conversion of prairie to Corn Belt," and the river was cleaner now than before European settlement.[44] (The Illinois River drains much of central Illinois and bits of Wisconsin and Indiana before emptying into the Mississippi River.) Winstanley was also quoted as saying that modern agriculture had "greatly cleansed" the Illinois River in the past half century.[45] He made other arguments against the idea that fertilizers were ending up in Illinois River and eventually the Gulf of Mexico.

To rebut those arguments, several scientists conducted many studies exploring the links between fertilizer use on land, nutrient levels in the Mississippi River and in its tributaries, and hypoxia in the northern Gulf of Mexico. The CENR reports alone had over forty different authors. But Rabalais and Turner were the most visible in the public eye. Rabalais especially was out in front, releasing to the press the latest estimates of hypoxia area in the Gulf, testifying before congressional committees, and enduring heckling by angry farmers. Even colleagues who sympathized with her and Turner thought they were going too fast and beyond the science. But in the end, they were vindicated: the data showing the link between fertilizer and nutrient levels were irrefutable. In 1999, Turner and Rabalais were given the $250,000 Blasker Award for Environmental Science and Engineering, and in 2012, Rabalais was awarded a MacArthur "genius" grant of $500,000. She earned her title, Queen of the Dead Zone.

Another outcome of Gulf dead-zone work by Rabalais and Turner and many others was the Harmful Algal Bloom and Hypoxia Research and

Control Act, first passed by the US Congress in 1998. Don Scavia, who was chair of the CENR Hypoxia Working Group at the time and worked on the language of the bill when he was at the National Oceanic and Atmospheric Administration, filled me in on the backstory[46] with help from Stephanie Bailenson.[47] Stephanie was a staffer for the US Senate Committee on Commerce, Science, and Transportation, which shepherded the bill into law. There are scientific reasons to put harmful algal blooms (HABs) and hypoxia together in one bill. HABs can cause hypoxia, and hypoxia can foment HABs; solving one problem would go a long way to solving the other. The political reason to have both in one legislation is even clearer. Senator Olympia Snowe, Republican from Maine, introduced a bill to examine HABs because of problems in coastal waters off her state and others. To ensure bipartisan support, the bill needed a Democrat co-sponsor. John Breaux, the senator we met in Chapter 3 who helped to kick-start research into the Gulf dead zone, was a Democrat and was very interested in Snowe's bill. But Breaux wanted to see work on the hypoxia problem, so that was added to the HAB bill. As a result, HAB went before hypoxia in the bill's title and got most of the money: $15.5 million vs $5 million for hypoxia.[48] The bill was reauthorized in 2004, again in 2013, and most recently in January 2019. Stephanie mentioned that it's one of the few bills that has been regularly reauthorized with steady bipartisan support.

Don Scavia had hoped the bill would do more than just fund research. A phrase in the first version of the bill, "Implementation Plan," implied something would be done to limit fertilizers and nutrients flowing into the Gulf, prompting some senators to put the bill on hold. It finally passed only after "Implementation Plan" was replaced with "Action Plan." In true Washington fashion, the first action was to set up a committee, a Task Force, to study the problem more and to "identify actions . . . to prevent, reduce, manage, mitigate, and control hypoxia and its environmental impacts." Even deciding on those actions was contentious. Don and others wanted to see TMDLs (total maximum daily loads) for nutrients and to follow the strategy paved by the US Clean Water Act for cleaning up the nation's rivers. The US Environmental Protection Agency used TMDLs to force states to build sewage treatment plants that brought oxygen and life back to rivers and estuaries like the Delaware. But TMDLs and the implied threat of restrictions on fertilizer use were too much for the Task Force members from the Corn Belt states. In response, Don suggested setting an "environmental endpoint goal" as he put it: reduce the dead-zone area to 5000 square kilometers, which is roughly 40 percent of the average area since 1985. He came up with the 5000 square kilometers target by extrapolating to the area of hypoxia before 1970 when nutrients hadn't yet reached today's high levels. The Corn Belt Task Force members and everyone else agreed to the area target, and that's been the goal ever since.

I wish that this story had a happy ending and that I could say the Gulf dead zone is shrinking in size to the 5000 square kilometer target. But that's not the case. If anything, hypoxic waters in the Gulf are expanding, and the dead-zone problem is getting worse.

Olympic Dead Zones in China

Agriculture is now the biggest source of nutrients feeding dead zones in most coastal waters around the world. Agriculture accounts for three to four times more nutrients entering into the Chesapeake Bay than does sewage.[49] For the Baltic Sea, nearly four times more nitrogen and fifty percent more phosphorus are from agriculture than from sewage.[50] The Black Sea is an interesting case, once thought to be a bright spot. The spread of hypoxia in the northwest shelf of the Black Sea during the 1960s and 1970 was thought to be caused by nutrient runoff from the collective farms in Ukraine, Romania, and other countries in the former USSR. Those farms could use state-subsidized fertilizers without the economic restraints faced by farmers in North America or Western Europe. With the fall of the Berlin Wall in 1989, the breakup of the USSR in 1991, and loss of the subsidies in the Black Sea watershed, fertilizer use declined by twofold to threefold in the late 1990s.[51] The author of an oft-cited *Scientific American* article stated that hypoxia in the northwest shelf of the Black Sea also declined and was gone by 2002. However, a more careful analysis didn't see a recovery and argued that the data were not good enough to reach a firm conclusion.[52] In any event, agriculture still contributes the lion's share of nutrients feeding the Black Sea.

The story of what causes dead zones in China is more complicated because the number of studies is small for the size and complexity of the country.

China's economic achievements over the last five decades are remarkable but so too are the country's environmental problems. China's meteoric rise has been accompanied by a shift from an agrarian society in the interior to an industrial one concentrated along the coast.[53] Augmented by migration from the countryside, the coastal population increased from 400 million in 1978 to 683 million and counting in 2018. The fraction of the country's GDP contributed by the coastal region, already high in 1978 (about 50 percent), is even higher today (60 percent). Along with skyscrapers and maglev trains, agricultural production has increased greatly over the last thirty years, spurred on by government subsidies to ensure the country could feed itself, in theory at least. Production was helped by "decollectivization" and other institutional reforms but even more by the adaptation of new technology,[54] similar to what we saw for the US Corn Belt: better seeds, mechanization, the application of herbicides and pesticides, and the heavy use of chemical fertilizers. One study calculated that 57 percent of the gain in agricultural production in China was due to technology, with fertilizer alone contributing 19 percent. In the mid-1970s for nitrogen and 1980 for phosphorus, China passed the United States in fertilizer use and now consumes far more than any other country.[55] Although China has only 9 percent of the planet's arable land, the country accounts for, according to 2013 data, 31 percent of nitrogen and 27 percent of phosphorus fertilizer used by the entire world.

The fertilizers do not stay put on a farm in Qingdao province any longer than they do on a farm in Iowa state. Even more than in the United States and Europe, fertilizers along with nutrients from wastewater and other urban sources end up in China's coastal waters. The efficiency of crops using nitrogen fertilizer ("nutrient use efficiency") is only 25 percent in China, half as good as in Europe (52 percent) or in the United States and Canada (68 percent).[56] Nutrients also come from manure, as more of the stuff has been dumped directly into rivers rather than spread on farmland when Chinese farmers switched to chemical fertilizers. One study estimated that direct manure discharge into rivers accounted for about 20 percent of total nitrogen inputs.[57] The consequences of this eutrophication of China's coastal waters have been dramatic.

The organizers of the Beijing Olympics had a problem on June 28, 2008, only four weeks before the games were to begin. Off the coast of picturesque Qingdao, a green alga blanketed 400 square kilometers of the Yellow Sea where the Game's sailing events were scheduled to take place.[58] The events were able to go forward only after one million tons of green slime were removed by more than 10,000 people at a cost of about 200 million RMB (USD $30.8 million). The Qingdao green slime was part of a massive green tide, which had started in the south, off Jiangsu province, scattered over about 84,000 square kilometers of the coastal sea. Satellite data dating back to 2004 indicate that the green tide first appeared in 2007 and has reoccurred every year since then, albeit not as extensively as the Olympic version. The green tide is made up of the green alga *Ulva prolifera*, commonly known as sea lettuce, unlike the microscopic algae and cyanobacteria floating over the hypoxic waters discussed so far. While sea lettuce differs from the microalgae in the Gulf of Mexico, the Chesapeake Bay, or the Baltic Sea, the reason why these algae grow so luxuriously is the same: excessive nutrients from agriculture, wastewater, and industry.

Excessive nutrients have led to hypoxia in at least two coastal regions of China, both south of Qingdao and Jiangsu province. The Pearl River and its estuary is surrounded by some of the largest cities in the world, including Hong Kong and Shenzhen, and has exceptionally high nutrient concentrations. The upper reaches of the estuary have experienced year-round hypoxia for many years; but even the lower part, previously thought to be too well-mixed to go hypoxic, is now seeing depletion of dissolved oxygen from its bottom waters.[59] The biggest dead zone in China is associated with the Changjiang (Yangtze) River, which flows past Shanghai on its way into the East China Sea. Reminiscent of the Mississippi River, the Changjiang contributes nutrients to a coastal ocean home to major fishing grounds. Nutrient inputs from the river have increased by a factor of ten since the 1960s, with the predictable effect on hypoxia. The Changjiang dead zone was 3000 to 4000 square kilometers when first studied in the late 1950s and 1960s, but it was nearly 16,000 square kilometers in 2009,[60] bigger than the Gulf of Mexico dead-zone average. More recent data indicate hypoxic waters off the Changjiang estuaries cover less area than the 2009 record, suggesting progress in limiting nutrients entering these coastal waters. But the Changjiang dead zone is still big. Because of leading the world in many economic areas, China also has world-leading dead zones.

Agriculture is not the only source of nutrients contaminating China's waterways. More nutrients come from sewage in China than seen in the United States and Europe, but that is changing. Domestic wastes account for about 39 percent of total nitrogen entering China's waterbodies, according to a 2019 study,[61] much higher than seen in the United States and Europe. But China has made great strides in treating those wastes. As of 2014, about 75 percent of urban wastes now go through a wastewater-treatment plant. These percentages vary greatly among the diverse regions of China. As wastewater treatment improves and fertilizer use increases, the contribution by agriculture to nutrient pollution in China's coastal waters is increasing. Today, overall, agriculture account for nearly 60 percent of all nitrogen contaminating the country's waters. As in the United States and Europe, agriculture is the main culprit causing China's dead zones.

6

Liebig's Law and Haber's Tragedy

My wife Ana loves to garden. She grows flowers along the sides of the house and vegetables in our small back yard. Gardening is not exactly my thing, but I try to help. I carry things too heavy for her to lift and trim branches from pear trees too high for her to reach. I'm often the one taking out the kitchen scraps to one of our four compost bins next to the back fence, and in fall, I rake leaves and then chop them up to layer with the kitchen scraps in the bins. Those kitchen scraps and leaves eventually turn into rich soil, which Ana spreads on the garden where needed. But there are times when it's not enough and she has to turn to chemicals. Miracle-Gro is the brand name of the fertilizer she uses. Here's what's in it:

Total Nitrogen	24%
3.5% Ammoniacal nitrogen	
20.5% Urea nitrogen	
Available phosphate	8%
Potash	16%
Boron	0.02%
Iron	0.15%
Manganese	0.05%
Molybdenum	0.0005%
Zinc	0.06%

This version of Miracle-Gro is called 24-8-16, referring to the amounts of nitrogen, phosphate, and potash. There are other versions for different plants, such as 18-24-16 for roses and 18-18-21 for tomatoes. Farmers with more than a backyard garden to tend may use something similar to Miracle-Gro with all of the nutrients needed by plants, or depending on their soil and crop, they may apply a fertilizer with only a few of the main ingredients. The soil may already have enough of the other nutrients for that particular crop.

It seems obvious that a farmer (or even my wife) should apply only the nutrient needed by the plants, if the others are already in excess. Plant growth would then reflect the added nutrient. As is often the case in science, this simple notion, that growth depends on the least-available nutrient, was not so obvious in the 19th century when the biology of plants was being discovered and the tools of organic chemistry were being worked out and applied to agriculture. The man who probably did the most in this area was Justus von Liebig, the namesake for Liebig's Law of the Minimum, which is what ecologists call the least-available nutrient idea. Liebig's Law dictates that plant growth depends on the one nutrient with the lowest concentration relative to plant needs. The problem is, Liebig didn't come up with the idea and was wrong about what nutrient was most important.

Liebig started his scientific life at the university in Giessen, Germany, a hamlet of only 5500 people in 1824.[1] There he perfected the analysis of plant and other organic material using laborious, tedious, and sometimes dangerous methods. Chemistry as a scientific field was emerging from the days when a substance was dissected by distillation and the products categorized as gas, phlegma, or residue. A precise balance was essential as were gasometric devices, and often the early chemist relied on his five senses. The Englishman Humphry Davy, discoverer of sodium and potassium and elected to the Royal Society in 1804 at the age of 26, relied heavily on his senses to make his advances in chemistry. He once tasted a mixture of water and "hot fermenting manure" of litter and cattle dung and discovered it had a "saline taste and disagreeable smell."[2] In Giessen, Liebig turned organic analysis into a science. Building on the work of others, he developed a combustion method to analyze organics and used the data to explore basic principles in organic chemistry. Among his early studies, Liebig determined the composition of an odd constituent in horse urine, hippuric acid ($C_9H_9NO_3$), and found the main acid (lactic) in sauerkraut. From these modest ingredients came modern organic chemistry.

While Liebig contributed to the fundamentals of organic chemistry, he also had many practical interests. He established in the early 1860s the Liebig Extract of Meat Company, which reduced boneless cattle cadavers into a mocha-colored, appetizing (so says his biographer) concentrate. Later that decade, he worked on another concentrate, this time of milk, sold through Liebig Registered Concentrated Milk Ltd. Less successful were his attempts to extract the antimalarial drug quinine from coal tar and to make mirrors without mercury. The eponymous Law of the Minimum comes from Liebig's book *Chemistry in its Application to Agriculture and Physiology*.[3] The first edition published in 1840 was followed by ten more editions in Germany, six in England, and nineteen in the United States. The first edition was only 195 pages long, whereas the definitive seventh one consisted of two volumes, clocking in at 1130 pages. Because of this book, Liebig could be called the father of fertilizer.

The book supplanted a similar one, titled *Elements of Agricultural Chemistry*, published in 1813 (1815 in New York) by Humphry Davy,[4] that dung-tasting Englishman, who recommended the application of manure to boost crop yields, supplemented by mineral fertilizers only when necessary. Davy's recommendation fit with a model of soil fertility, the humus theory, that was popular in the early 1800s. According to the theory, plants live by absorbing through their roots humus, the dark, moist organic stuff of soils. The theory held that plants obtained most of their carbon from humus even though it had been known since the late 1700s that photosynthesizing plants use carbon dioxide from the atmosphere. An important experiment refuting the humus theory showed that some plants could be grown in sand and water without any humus. Building on that result and using the methods developed by Liebig, the German Karl Sprengel carried out a series of analyses to further disprove the humus theory and to demonstrate that humus is valuable, not because of its carbon, but because of its minerals or inorganic nutrients like nitrate, phosphate, and the potassium salts that make up potash. He went on to show in 1838 that plant growth depended on the nutrient in shortest supply.[5] So, Liebig's Law of the Minimum really should be called Sprengel's Law of the Minimum. Nearly two centuries later, Sprengel was finally recognized for his contributions, although he still shares the spotlight with Liebig. Each year the Association of German Agricultural Experimental Stations awards the Sprengel-Liebig Medal for outstanding service or achievement.

Although Liebig gave Sprengel his due credit initially, he has overshadowed Sprengel because he was a better publicist and more prolific. There were all those editions of *Agricultural Chemistry*, the hundreds of scientific papers he published (thirty per year on average between 1830 and 1840), the thousands of letters he mailed off to colleagues and competitors, and the countless speeches and lectures he gave. Early agricultural chemists may not have given Sprengel more credit because he believed that humus was the best fertilizer, in contrast to Liebig who argued for mineral fertilizers. Liebig was right about the mineral theory, that plants use mineral forms of carbon, nitrogen, phosphorus, potassium, and the other elements needed as nutrients.

But he was wrong about which mineral fertilizer was most important. Before discussing how he was wrong, I need to head off the ire of organic gardeners (and my wife): organic fertilizers have many virtues. They slowly release mineral nutrients and leave behind humus that improves soil quality. Soils high in humus are deemed rich, and rightly so. But plants use inorganic chemicals like nitrate and phosphate, instead of directly taking up the organic chemicals containing nitrogen and phosphorus.

Liebig was so sure of his views that he lent his name to companies in Britain and Germany to make mineral fertilizers. The German enterprise made six types that were designed for different crops using different mixes of the ashes of plants and bones, gypsum (calcium sulfate), potassium silicate, magnesium sulfate, and crucially, only a bit of ammonium phosphate. Miracle-Gro it was not, but Liebig's fertilizer supplied enough of the two main ingredients making up a modern fertilizer: potassium and phosphorus. The potassium was mostly from the plant ashes, whereas today potassium-rich potash in Miracle-Gro and other modern fertilizers is now mined from deposits mostly in Canada, Russia, and Belarus.[6] A second main ingredient, phosphorus, came from the other ash, that of bones. Roasting bones at a high temperature removes the protein and other organic components, leaving behind ceramic minerals such as hydroxyapatite, rich in phosphorus. Liebig wasn't the only person using bones for fertilizer. In the mid-19th century, the demand in Great Britain for bones outstripped the supply from knacker's yards and abattoirs, forcing importation from other countries, reportedly including mummified cats from Egyptian pyramids and remains from the battles fought at Leipzig, Waterloo, and the Crimea.[7] Liebig claimed, "Great Britain is like a ghoul, searching the continents for bones to feed its agriculture."[8]

Today, phosphorus comes from deposits of rock phosphate, phosphorite. The largest deposits are in Morocco and the Western Sahara, although China is the world leader in mining phosphorite.[9] The United States has nearly all the phosphorus it needs, mostly from phosphorite mines in Florida and North Carolina that provide 75 percent of the phosphorus used by the country. In 2017, the United States imported only 6 percent of the total phosphorus used for fertilizers, detergent, and other applications.

Liebig was so confident about the efficacy of his fertilizers that he didn't bother to test them with real crops, not even on Liebig Heights, the ten acres of land he owned outside of Giessen. He had plans for a summer home surrounded by an English garden. Accompanied by glowing pamphlets, the patented fertilizers went on the market in the fall of 1845. Liebig's fertilizers turned out to be a fiasco. After finding out the hard way that Liebig's fertilizers didn't work, farmers vilified him, and the companies went bust, forcing Liebig by 1849 to sell Liebig Heights.

The fertilizers had several problems. It didn't help that Liebig had the fertilizer ingredients baked together into insoluble clumps to slow diffusion away from the plants. The cooked fertilizer stayed on the soil surface unless ploughed in, another expense added to an already high price. Solving that problem, however, didn't fix the fertilizer's main flaw: its low nitrogen content. While Liebig's fertilizer mixes had enough of two of three key elements, phosphorus and potassium, they didn't have enough of the third, nitrogen. He felt the mixes didn't need much because he thought crop plants could get enough nitrogen from the ammonia in rainwater. He and others found the chemical in rain, and even though concentrations were very low, Liebig calculated that a field of average size in his day could receive 88 pounds of ammonia per year, more than enough for a crop of corn or of any grain. But that calculation was wrong—"dodgy" is the word used by Liebig's biographer. Crops need much more nitrogen than what rain can provide. That became known just about the time Liebig's fertilizers went on the market. Working at the Rothamsted experimental farms, which are still in operation today, Sir John Benet Lawes showed that nitrogen fertilizers were most needed to increase crop yields.[10] Liebig wasn't convinced about the importance of nitrogen fertilizer until it was too late.

Liebig didn't have many options to get nitrogen into his fertilizer mix, especially if he kept to only mineral forms of the element. He and everyone else in the 19th century and the early 1900s didn't know how to convert or "fix" the most abundant form of nitrogen, nitrogen gas in the atmosphere, into something useable by plants. They were forced to use natural deposits of fixed nitrogen: nongaseous chemicals like nitrate, ammonium, and organic-bound nitrogen. Looking at Liebig's options and the history of nitrogen fertilizers is a bit of a detour, but worth it. It helps to explain why dead zones could not have proliferated much earlier than the mid-20th century, and the history has tales of both human ingenuity and one man's tragedy.

For War or Agriculture

Liebig knew about guano,[11] the most common nitrogen fertilizer used in the 19th century. He didn't think it was necessary to include this in his fertilizer, and it didn't help that guano was expensive, being imported from Peru.[12] Bird droppings, the stuff of guano, are everywhere, but guano deposits were especially huge along the coast of Peru where large flocks of nesting sea birds feed on abundant fish rich in protein. The dry climate ensured that bird droppings weren't washed away and could accumulate over millennia. Guano was used in coastal areas by indigenous South Americans for centuries and later by the Incas for highland crops in the Andes. The Spanish conquistadors knew of guano's power as a fertilizer, but it wasn't exported to Europe or North America until around the time Liebig was promoting his nitrogen-deficient fertilizer. Guano quickly came to dominate the commercial nitrogen fertilizer market, rocketing from nothing in the early 1840s to an annual peak of 600,000 tons in the late 1860s. But the deposits were too small to last many years of aggressive extraction. By the opening years of the 20th century, guano from South America supplied less than 1.5 percent of global nitrogen needs. Today organic gardeners, not including my wife, as far as I know, still prize guano, albeit from fruit-eating bats. One product I found on the web is Dr. Earth 726 Premium Bat Guano, said to be "Trubiotic, the life and intelligence inside."

Another nitrogen source for Liebig's fertilizers could have been sodium nitrate. In the 19th century, it also came mainly from South America, this time Chile. Nitrate-rich rocks, or caliches, are found inland about 25 to 150 kilometers from the coast, high in a sere plateau between the Pacific Coastal Range and the western slope of the Cordillera de Los Andes. How the nitrogen-rich deposits got there is not totally known, but one idea is that sodium nitrate and other salts are the remains of an ancient inland sea that had dried up. The salts were preserved in the arid, lifeless plateau, untouched by rain or plants. The nitrogen content of caliches ranges from 6.5 percent to as much as 70 percent in some *salitreras* (nitrate fields in English) in Tarapacá. Exports to Europe and North America began earlier than the guano market and lasted longer, until World War I. But like guano, nitrate deposits in Chile were too small to satisfy the world's nitrogen needs for long. The total amount of nitrogen in caliches, estimated to be 920 metric tons in 1908, was less than the amount of ammonia made by today's industrial process in two years.

That the world was running out of nitrogen alarmed scientists and leaders of the Great Powers in the closing years of the 19th century. In 1898, Sir William Crookes stated in his presentation to the British Association for the Advancement of Science, "England and all civilised nations stand in deadly peril of not having enough to eat" because the supply of nitrogen fertilizer would soon become inadequate.[13] New sources of nitrogen were needed for giving life, but also for taking it away. Along with sulfur and charcoal, sodium nitrate is one of the three main ingredients of gunpowder. (American readers of a certain age may remember that the Oklahoma City bomber used ammonium nitrate fertilizer to blow up the Alfred P. Murrah Federal Building in 1995, destroying or damaging hundreds of buildings over 16 city blocks and killing 168 people.) In 1900, about half of the nitrate imported by the United States went into explosives. But at the beginning of the 20th century, the world's largest importer of nitrate was Germany.

Even if caliche deposits were unlimited, Germany had many reasons to find another nitrogen source. Its civilian and military leaders were acutely aware that the country's supply of Chilean nitrate could be easily cut off by a naval blockade, which would starve the German war machine quicker than its people. Germany wasn't the only country worried about running out of nitrogen, whether for explosives or fertilizer. The early years of the 20th century marked the beginning of the race among nations to make nitrogen-rich chemicals by industrial means. These attempts included collecting the ammonia produced during pig iron smelting and transforming nitrogen gas into cyanamide (CN_2H_2) or nitrate. Although nitrogen-fixing factories were built using these early approaches, the output was too low to supply enough nitrogen for war or agriculture.

The approach that solved the nitrogen problem came from two Germans.[14] In 1909, Fritz Haber discovered that ammonia could be made by combining nitrogen gas and hydrogen gas in the presence of a metal catalyst under high heat and pressure. Carl Bosch turned Haber's small benchtop apparatuses into mammoth factories. The first using the Haber-Bosch process started to produce ammonia on May 18, 1910. Two months later, it could make enough ammonia to fill up only a five-kilogram container, but then production ramped up incredibly fast. By 1923, a two-kilometer long plant near the small village of Leuna, Germany, was capable of producing 200,000 tons of ammonia in a year. The rapid expansion of the nitrogen-producing factories was driven by the Great War; but once it and its last battle (World War II) finally ended, the factories in Germany and elsewhere could turn from ammunition to fertilizer.

Today, without the Haber-Bosch process, Vaclav Smil has calculated that the world would have three billion fewer people. Without the Haber-Bosch process, he argues, agriculture could support only about half of today's global population with "a very basic, overwhelmingly vegetarian diet."[15]

The Haber-Bosch process is the reason why Sir William's dire prediction has not come to pass. Yes, malnutrition is still too common even in "civilized" countries, as is starvation in poor, war-torn, or climate change-stressed regions: but those problems are not due to the lack of fertilizer. Less nitrogen would be needed if people, especially in the United States, ate less nitrogen-rich food, aka meat, a topic I'll return to in Chapter 10. However, even if people chowed down fewer hamburgers and steaks, the Haber-Bosch process would still be needed to feed the billions of people around the world. Today the amount of nitrogen produced by the Haber-Bosch process is huge, roughly the same as the amount from natural processes. To say it in another way, we are doubling the input of fixed nitrogen into the biosphere. Depending on your diet, about half of the nitrogen in your body came from a factory using the Haber-Bosch process.

Haber was awarded the Nobel Prize for Chemistry in 1918, while Bosch had to wait until 1931 to get his. That's the human ingenuity part of this detour.

The tragic part is Fritz Haber's life. If not tragic, it was certainly full of contradictions. His discovery eventually would better the lives of billions, but much of the nitrogen first made by Haber-Bosch factories went into ammunitions that killed millions during World War I. More damning are Haber's other contributions to the German military (Fig. 6.1). Far from being an academician secluded in an ivory tower, Haber was vocal in supporting the nationalist aims of a militant Germany and active in the war effort as soon as the battles began.[16] As he put it, scientists should be "for humanity in time of peace, for the fatherland in time of war."[17] Along with many other prominent scientists and intellectuals, he signed a letter in September 1914 supporting Germany's declaration of war. After the start of the conflict, he tried to enlist in the army but was too old. Instead, he was made head of the Ministry of War's Chemistry Section where he worked on chlorine gas and other weapons of chemical warfare. He even ventured onto the battlefield to gauge chlorine's effectiveness, first on Allied troops during the Second Battle of Ypres in spring of 1915 and then later on Russian troops at the Eastern Front. When questioned after the war about the legality and morality of chemical warfare, he argued that the French had used gas first and that it was no more gruesome than a bullet or bayonet.

With the end of the war, Haber continued to work for the fatherland but also for humanity—his colleagues and the German people. In November 1918, Haber was asked to take over the chemistry section of the National

Figure 6.1 Fritz Haber in 1916. He was so appreciated by the Kaiser that he was made captain, a rank he kept throughout the war. From Science Photo Library.

Office for Economic Demobilization, which was charged with redirecting the chemical industry of Germany into peacetime work and away from warfare. However, he lasted only until February 1919, as he thought it better to concentrate on his institute, the Kaiser Wilhelm Institute of Physical Chemistry. He was adroit at securing financial support for the institute and skillful in managing the senior scientists and nurturing the junior ones. He made the institute into one of the best in physical chemistry for its time. One of the institute's notable projects was the transformation of warfare chemicals, which Haber had developed, into chemicals for controlling pests that threatened limited food supplies. One of his postwar projects has an oceanographic connection. In one of his efforts to improve Germany's dire financial state in the 1920s, Haber worked on a scheme to extract gold from the oceans after he learned about the precious metal being in seawater from the Swedish chemist Svante Arrhenius. His initial efforts were promising. In the summer of 1923, Haber embarked on the passenger ship *Hansa* to try out methods in the North Atlantic Ocean while the ship sailed to New York, and then later that year, he cruised to Buenos Aires to see if the warmer waters of the South Atlantic had higher gold concentrations. Eventually, Haber was forced to give up. He discovered what oceanographers are all too aware of. As he put it,

> There is nothing as varied as the conditions in the oceans. Perhaps one day, somewhere, a place will show up where gold may be found, where precious metals accumulate. . . . I have given up looking for this dubious needle in a haystack.[18]

The oceans do have gold but at concentrations too low to be extracted economically.

In 1932, Haber had long planned a two month stay in the French Riviera to recuperate and regather his strength. In a letter to his medical advisor, he mentioned that putting pressure on his neck artery was so useful for cutting off "light attacks" that he no longer needed nitroglycerin.[19] He was besieged by insomnia, financial demands from his divorced wife (his second—his first had committed suicide), and mounting concern about the future. Near the end of Haber's stay in France, on January 30, 1933, Adolf Hitler became chancellor of Germany. Although aware of the anti-Semitic views of the National Socialists, Haber didn't fully appreciate how fast dangers facing his institute and him would arrive. The Law for the Restoration of the Professional Civil Service, passed on April 7, 1933, forced the "retirement" of all non-Aryans, especially Jews. Haber attempted to delay implementation of the law at his institute and wrote a flood of letters to friends abroad in order to find positions for Jewish colleagues and co-workers. Soon, however, the law claimed him as a victim. In spite of being an ardent supporter of the fatherland and having converted to Lutheranism at an early age, Haber lost his position at his own institute because he was a Jew. He was forced to leave Germany, traveling first to Paris to visit his son and then to Switzerland where he met Chaim Weizmann, who later would become the first president of Israel. After a brief stay in England, Haber agreed to Weizmann's offer to become the director of the Daniel Sieff Research Institute in Palestine, now the Weizmann Institute in Israel. But before he could make it to the Middle East, on January 29, 1934,

Haber died in a Basel hotel room at the age of 65. Members of his extended family were less fortunate. Many were killed during the Holocaust by the gases Haber's institute had developed.

Haber's story has taken us a long way from hypoxic bottom waters. To segue back to the science, I'd like to note the parallels between Haber the man, the Haber-Bosch process, and nitrogen the element. Just as some chapters of Haber's life are admirable, whereas others are deplorable, the chemical process he invented is immensely beneficial to society: yet it also provides the raw material for manufacturing bullets and bombs. Likewise, nitrogen is essential for all life on Earth, yet too much of it now threatens the biosphere.

How to Start a Dead Zone

Liebig's biographer wrote, "From a twentieth century perspective, nearly everything in *Animal Chemistry* [his multi-edition textbook] was wrong."[20] Liebig was wrong about the structure and composition of proteins and about how we digest vegetables and meat, and his bile theory "proved to be complete invention." Most important here, his fertilizers were certainly wrong, spectacularly so. Yet his general idea about plants using inorganic nutrients was right. He was also right about his (and Sprengel's) Law of the Minimum. Although plants and algae need several nutrients, their growth depends on the one in the shortest supply. Fast forward a century and a half, aquatic ecologists were thinking about the Law and which nutrient sets rates of algal growth in coastal waters like the Gulf of Mexico. The answer to that question is also the answer to which fertilizer causes algal blooms and ultimately hypoxia in bottom waters. Perhaps we need to stop only one fertilizer to solve the dead-zone problem.

One aquatic ecologist was relieved to learn about the Law of the Minimum: an evil ecologist who wants to start a dead zone for unknown sinister reasons. He had thought about using Miracle-Gro. Algae need the same nutrients, more or less, as plants on land, so Miracle-Gro would work. But then the evil aquatic ecologist calculates that to get an algal bloom the size of the Gulf dead zone using Miracle-Gro 24-8-16 at retail prices, he would have to shell out nearly a trillion dollars. Even though he is filthy rich, he doesn't have that kind of money. Fortunately for him, most of the eight elements in Miracle-Gro would be wasted in most oceans and seas, so he doesn't need those. Certainly, he could leave out the zinc, manganese, and boron, as there is enough of those in marine waters. Thirty years ago, studies suggested that growth of important cyanobacteria is limited by molybdenum, but then it was found that seawater has enough of that element too. Iron is limiting in a few oceanic regions like the Equatorial Pacific, far from land and sources of iron-rich dust; but the Gulf of Mexico and the other regions with hypoxia problems receive more than enough iron brought in from nearby land by wind and rivers. Even potash, the 16 of Miracle-Gro 24-8-16, the version of the fertilizer used by my wife, isn't needed to start an algal bloom. Salt water has enough salts of potassium. No, the evil ecologist can focus only on the 24 and the 8: the nitrogen and phosphorus.

Now the story becomes more complicated. If serious about getting a big bloom and a large dead zone at the lowest price, our evil ecologist should find out which type of nutrient, nitrogen or phosphorus, limits algal growth in his targeted region. Liebig's Law rules. Of course, identifying the limiting element would not only help the evil ecologist: it would also help us find the most cost-effective way to shrink dead zones, if not eliminate them entirely.

Geochemists concentrate on phosphorus. In their influential book on nutrient cycling, W. S. Broecker and T.-H. Peng called phosphate the "ultimate limiting nutrient."[21] Phosphate, the most common mineral compound with the element phosphorus, has a slow and arduous journey to the oceans, starting with erosion from rocks on land, followed by transport by rivers to the sea. In contrast, there is lots of nitrogen in air, 70 percent of the atmosphere being nitrogen gas, and there is a mechanism, nitrogen fixation, which turns nitrogen gas into a form, ammonia, useable by plants and algae. Broecker and Peng argued that if nitrogen became limiting, nature would favor the growth of nitrogen-fixing organisms. That growth would eventually alleviate the nitrogen deficit and return the oceans to phosphorus limitation. Like Liebig's opinion about which fertilizer was most needed by land plants, geochemists believe phosphorus is the "most indispensable nourishment" for algae in the oceans.

Biologists have a problem with the notion of nitrogen fixation being fast enough to provide enough useable nitrogen for algal growth. Sure, there is lots of nitrogen gas in the atmosphere and water, but nitrogen fixation is not easy nor that common. The Haber-Bosch process requires lots of heat and energy and tons of pressure. Microbes are able to carry out the same reaction at room temperature and atmospheric pressure, but they also require lots of energy and a big, complex enzyme, nitrogenase. The challenge is to break

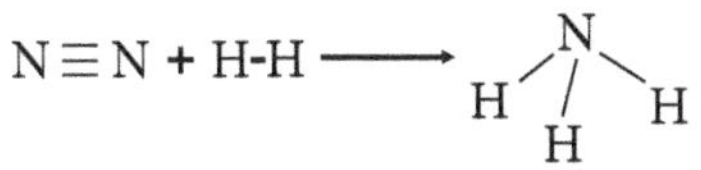

$$N \equiv N + H\text{-}H \longrightarrow \quad H^{\diagup\hspace{-0.3em}\underset{H}{\overset{N}{\mid}}\hspace{-0.3em}\diagdown} H$$

Figure 6.2 The breaking apart of nitrogen gas (N_2) by hydrogen gas (H_2) to form ammonia (NH_3), the reaction behind "nitrogen fixation" by both microorganisms and the Haber-Bosch process.

apart the three bonds making up dinitrogen gas (N_2) and to convince the nitrogen atoms to pair up with hydrogen (Fig. 6.2).

Few organisms can do all that. Besides humans and their factories, the only organisms capable of nitrogen fixation are a handful of bacteria, such as those molybdenum-needy cyanobacteria. It's not done by the most common cyanobacterium in the oceans, *Prochlorococcus*. Even those microbes that do fix nitrogen do it very slowly. The nitrogen fixation rate is usually just a tenth of the rates for other nitrogen processes.

While it's very hard to turn nitrogen gas into a fixed state like ammonia, turning ammonia back to a gas is fairly easy and relatively fast. There are two main parts to the conversion, carried out by very different microbes. The first part, the conversion of ammonia to nitrate or "nitrification," is important here because it consumes dissolved oxygen. Nitrifying microbes oxidize ammonia while also depleting oxygen, analogous to what we and other animals do with organic chemicals. Instead of organics, nitrifying microbes get their energy from ammonia. The huge difference is that ammonia oxidation doesn't yield much energy, less than a tenth of the energy produced when heterotrophs oxidize organics. That's why nitrification is carried out by only a few bacteria and archaea. However, nitrification can use up a lot of dissolved oxygen and contributes to forming a dead zone. It is especially important downstream of wastewater-treatment plants that remove only organic material and release large amounts of ammonia.

The second part of turning fixed nitrogen to nitrogen gas starts with nitrate in a process called "denitrification." The denitrification reaction is very common among bacteria and even some protozoa, which use nitrate to oxidize organic material in order to obtain energy for cellular processes. Here, nitrate is taking the place of dissolved oxygen. Denitrifying organisms would prefer to use oxygen to oxidize organic material because it generates more energy. But if they don't have enough oxygen, they can turn to nitrate. Although there is an energetic advantage in using oxygen instead of nitrate, the advantage is rather small, only about 100 kilojoules per molecule, less than 5 percent of the total energy gained by using oxygen.[22] (Joule has replaced calorie as the term preferred by scientists when discussing energy.) The energetic advantage is enough for organisms to favor oxygen over nitrate, but it is not a big enough drop to slow down organic matter oxidation.

In the end, the synthesis of useable nitrogen chemicals like ammonia and nitrate is slow, whereas the loss of useable nitrogen is fast. The net result is low concentrations of nitrate and other forms of useable nitrogen in the surface layer of aquatic habitats where algae need the element for growth.

So, who is right about the most important element? The geochemists who favor phosphorus or the biologists who think the answer is nitrogen? One way to find out is simply to add one element or the other (more precisely, a nutrient with the element) to water and see which one works. The experiment has been done in the Experimental Lakes Area, in Manitoba, Canada, as part of a decades-long program to explore many topics in aquatic ecology and limnology. Entire lakes have been dosed with nitrate or phosphate, and the effect on algae, oxygen, and many other environmental properties were monitored over time. The work demonstrated that algal growth in these Canadian lakes is set by the phosphorus supply, and usually that's the answer for lakes around the world.[23] More phosphate results in more algae and less oxygen in bottom waters. A victory for the geochemists. Results from these experiments and others were instrumental in convincing governments around the world to ban phosphate in detergents, with the goal of reducing excessive nutrients, eutrophication, in lakes. Eutrophication causes damaging algal or cyanobacterial blooms that disrupt food chains, give water a foul smell and a dirty-sock taste, and ultimately deplete oxygen in bottom waters. These problems are caused by excessive phosphorus, the element setting the rate of algal growth in lakes.

With one exception, experiments analogous to adding nutrients to an entire lake can't be done in coastal waters or the open oceans. The exception is iron. One piece of evidence indicating iron limitation came from experiments in which 450 to nearly 3000 kilograms of iron was dumped into the ocean from ships sailing over tens to hundreds of square kilometers.[24] In places like the equatorial Pacific and the Southern Ocean, the added iron stimulated algae to grow faster. Along with injecting sulfur dioxide into the atmosphere to promote cloud formation, iron fertilization is often mentioned as a geoengineering solution to global warming because the iron-enhanced algal growth could pull down more carbon dioxide from the atmosphere. Algae don't have a lot of iron, so relatively little is needed to start a bloom in waters limited by that element. But the comparable experiment can't be done to test for nitrogen versus phosphorus limitation. It's not possible to add enough of either to change algal growth in coastal waters, even if it were environmentally sound and legal.

Rather, experiments testing nutrient limitation are done in small bottles or tanks. One of the biggest experiments was done with 13,000-liter tanks at the Marine Ecosystem Research Laboratory (MERL) on the shores of Narragansett Bay in Rhode Island (Fig. 6.3). As with the whole-lake experiments, phosphate or nitrate or both was added to the MERL tanks, and algae and oxygen were followed over nine weeks.[25] Unlike the lake experiments, algae grew the most and oxygen production was highest when nitrate was added to the seawater. Biologists won, at least in Rhode Island.

What about dead zones like the one in the Gulf of Mexico? What governs algal growth in surface waters above hypoxic bottom waters? It's in these

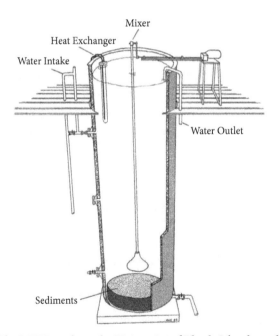

Figure 6.3 The MERL tanks at the University of Rhode Island, used to test which nutrient limits algal production in marine waters. The tanks are 1.8 meters in diameter and 5.5 meters deep. The facility had 14 of these tanks when it was in full operation. Today only four are available along with ten smaller ones.

surface waters that algae grow too exuberantly, creating too much organic material that sucks oxygen out of bottom waters. The surface layer is
mostly marine, so perhaps these waters are limited by nitrogen. Yet, rivers
flow into them, carrying two key dead-zone ingredients: fresh water and
nutrients. Because algal growth in fresh waters is supposed to be limited
by phosphorus, perhaps algal growth in the dead zone surface layer is also
phosphorus-limited.

To examine that question, scientists have done enrichment or bioassay
experiments like the MERL work, albeit with Coke bottle-sized containers,
not 13,000-liter tanks. These experiments revealed that nitrogen is usually
the limiting element for the Baltic Sea, Chesapeake Bay, the Black Sea, and
the northern Gulf of Mexico. I have to mention the 170 experiments done
by Timo Tamminen in the Baltic.[26] Yes, phosphorus seems in short supply
in low-salinity water of the Baltic and the upper Chesapeake Bay in spring.
But nitrogen controls algal growth in most of the Baltic (the Baltic Proper)
and the Chesapeake, especially in the surface layer above the bottom waters
most depleted in oxygen. Nitrogen is also the answer for most of the Gulf
of Mexico. That nitrogen is the limiting nutrient for the Gulf won't surprise
you if you remember the tight relationship between nitrogen fertilizer use in
the Mississippi River basin and dead-zone area in the northern Gulf, discovered by Gene Turner and Nancy Rabalais. The husband-wife duo also noted a
tight connection between nitrate and algal biomass. Even before making that
connection, they started to do nutrient addition experiments to see what limited algal growth, eventually completing 158 experiments over thirty years.[27]
They usually saw nitrogen limitation and only occasionally phosphorus limitation in the northern Gulf.

There were times, however, when adding nitrate or phosphate didn't do
anything. One explanation is that the algae aren't receiving enough light.
Light limitation explains why fertilizer leaching from Corn Belt states doesn't
turn the Mississippi River green. It also explains how the fertilizer can make
it all the way to the Gulf without being sucked up by algae and rooted plants
along the way. The Mississippi is brown with sediment, its high turbidity
stopping the light necessary for algal growth. Algae certainly grow in the
Mississippi, but not enough to deplete the nutrients. More than enough fertilizer leaching from an Iowan cornfield survives the thousand-mile journey
down the river to the Gulf where it fuels algal blooms and oxygen depletion.

However, there is more to the story, parts not told by the bioassay
experiments. There are problems with these experiments and complications

to the simple idea of nitrogen controlling algal growth. Ecologists are nervous about extrapolating from even a 13,000-liter tank to the billions of liters of water in a sea or coastal region. Even if the tank perfectly mimics the ocean, ecologists can't do enough of the experiments to cover all times and places within a region suffering from hypoxia. Fortunately, there is another approach that doesn't need a bottle or tank. We can look at the ratio of the nitrogen to the phosphorus nutrients and then see how that ratio compares with the Redfield ratio. The ratio and the man behind it are discussed next in Chapter 7.

7

The Case for Phosphorus

When we last left the evil ecologist in Chapter 6, he was trying to make a dead zone by stimulating an algal bloom. After reading about the enrichment experiments, he thinks that adding just nitrogen fertilizer without phosphorus would be enough to start the bloom and get his dead zone. The bioassays found that the addition of nitrogen nutrients was enough to stimulate algal growth in coastal waters, implying nitrogen fertilizer would be enough to create a dead zone. But then the evil ecologist wonders if he should trust results from a few small bottles. Even though some of the studies had done hundreds of experiments, they couldn't cover an entire coastal region all of the time. Because the evil ecologist can't afford a mistake, he wonders if there is another way to see whether he should add nitrogen or phosphorus to start a dead zone.

Of course, the rest of us are more interested in stopping dead zones, or at least shrinking them to a preindustrial agriculture size, 5000 square kilometers in the case of the Gulf of Mexico. As will be discussed in Chapter 10, some solutions would work regardless of whether the limiting element is nitrogen or phosphorus. Vegetated buffer zones, for example, suck up fertilizer leaching from a field and stop both nitrogen and phosphorus nutrients from entering streams and eventually coastal waters susceptible to hypoxia. But for many of the other solutions, to be cost-effective, we need to know whether to concentrate on one or the other element. Nitrogen and phosphorus differ in their fate in the environment, how wastewater-treatment plants remove them, and the amounts needed by crops and ending up in manure. So, in addition to the evil ecologist, ecologists with better intentions and ecosystem managers very much care whether nitrogen or phosphorus sets the dimensions of a dead zone.

Given the stakes, we need to get the answer right. That justifies having another approach to corroborate the bioassay experiments. One alternative looks at nutrient concentrations, calculates the ratio of phosphorus to nitrogen for the nutrients, and then compares that ratio to a standard: the Redfield Ratio.

Allie's Ratio

Alfred Clarence Redfield, "Allie" to family and friends, became interested in the oceans while summering as a kid in Woods Hole and Barnstable on Cape Cod (Fig. 7.1).[1] He had no formal training in the marine sciences or much else to prepare him for the work that would eventually lead to his eponymous ratio. For someone so famous among oceanographers, he published only a handful of papers in the field, among a few other publications on an eclectic mix of topics. His education was equally eclectic and unusual by today's standards. He was supposed to be educated as a "museum man" and had only introductory courses in chemistry and a bit of physics at Harvard University. After graduating, Redfield went on to do a PhD at Harvard in animal physiology, finishing in 1917, and then he became an assistant professor at the Harvard Medical School. There he worked on the blood of squid, horseshoe crabs, and other invertebrates found on Cape Cod beaches, but most of his scientific oeuvre is about the respiration of porpoises, sea turtles,

Figure 7.1   Alfred Redfield in 1955. J. Hahn, Copyright Woods Hole Oceanographic Institution.

sea lions, and other animals, including humans. In 1930, he was recruited to the newly formed Woods Hole Oceanographic Institute in spite of having no background in oceanography. Although he still had a position at Harvard, first as a full professor in 1931, then a few years later as the director of the university's Biological Laboratories, he felt "rather intangible pressure to do work which was more specifically oceanographic." So he "spent a winter making a survey of the Gulf of Maine, of the chemistry and physics of water." It was a winter well spent.

Redfield looked at nitrate and phosphate concentrations in the water and the nitrogen and phosphorus content of the plankton, the microscopic menagerie of free-floating algae and animal-like creatures at the base of a food chain that makes all other life in the oceans possible. Redfield found that although the amount of nitrogen and phosphorus in the two nutrients and in the plankton varied greatly, the ratio of the two was surprisingly the same. In the Gulf of Maine and in the other oceanic regions with nutrient data available at the time, the ratio of nitrate to phosphate was what we now call the Redfield ratio: 16:1. The nitrogen to phosphorus of the plankton was also 16:1 everywhere he looked. That's astounding. As he put it, "the relative quantities of nitrate and phosphate occurring in the oceans of the world are just those which are required for the composition of the animals and plants which live in the sea."[2] It wasn't a coincidence. Rather than being set by abiotic, physical forces, the ratio of nutrients was determined by the organisms and their biochemical make-up: the relative amounts of protein, nucleic acids, and other macromolecules with varying numbers of nitrogen and phosphorus atoms. Why precisely 16:1 is a bit hard to explain, but it is clear why microbes in the plankton have more nitrogen than phosphorus. Microbes, plants, and animals need more nitrogen because they have more protein and other nitrogen-rich macromolecules than chemicals with lots of phosphate. The relative amount of the macromolecules in turn is set by evolution to maximize growth and reproduction of marine planktonic organisms.

Peter J. le B. Williams has argued that Redfield's observation foreshadows by several decades the Gaia hypothesis developed by James Lovelock and Lynn Margulis in the early 1970s. Gaia is the primordial goddess of Earth in ancient Greek mythology. Lovelock and Margulis hypothesized that organisms regulate their abiotic surroundings to perpetuate livable conditions on the planet. It was once thought that a Gaia-ruled biosphere would withstand climate change even if it meant expelling its dominant inhabitant, *Homo sapiens*. Some have mused that the Covid-19 pandemic is a "symptom of how sick of us is Gaia, our planet."[3] That would be too much for most ecologists to accept. Even if the strong, nearly mystical form of the hypothesis is no longer believed,[4] it remains true that there are many interactions and feedbacks between organisms and their environment. Organisms do shape their surroundings, and the abiotic parts of the planet would look very different without biota. In any case, although Redfield never called forth Greek goddesses, the title of one of his most famous papers has a Gaia-like flavor: "Biological control of chemical factors in the environment."[5] Gaia or not, the Redfield ratio is a profound statement about how marine organisms shape and are shaped by their environment.

Soon after Redfield came up with his ratio, biological oceanographers realized it could be used to explore nutrient limitation of algal growth; when nutrients aren't Redfieldian, algal growth might be limited by the nutrient in short supply. A nutrient ratio above the Redfield ratio of 16:1 indicates the water has lots of nitrogen and not enough phosphorus, making phosphorus the limiting element, the one that sets how fast algae grow. Below 16:1, nitrogen is in short supply and is limiting. Because it is much easier to measure nutrient levels than to do a bioassay experiment, the Redfield approach makes it possible to look at potential limitation over a much wider swath of a region over a much longer time frame.

Phosphorus Limitation in the Gulf?

We can now use the Redfield ratio to revisit the question of which nutrient, nitrogen or phosphorus, sets algal growth in surface waters above dead zones and thus is most important governing the extent of hypoxia. We'll first look again at the Gulf of Mexico. Remember the bioassays had indicated that the northern Gulf is mainly limited by nitrogen. As discussed in Chapter 5, early experimental work and other data led the National Science and Technology Council's Committee on Environment and Natural Resources (CENR) to recommend in 2000 that controlling nitrogen inputs was the way forward to solving the Gulf's dead-zone problem. Initial studies using nutrient ratios came to a radically different conclusion.

One of the most influential reports using nutrient ratios to understand the Gulf dead zone has never been officially published. Drafts of the report written by scientists in the Atlanta office of the US Environmental Protection Agency (EPA) were given to me by Don Boesch. We last saw Don in Chapter 4 dueling with Tom Bianchi about alternative sources of organic material fueling oxygen use in the northern Gulf. Now Don and others were fighting against the samizdat EPA report that argued for phosphorus limitation.[6] The EPA report pointed out that the nitrogen to phosphorus ratio for the Mississippi River is very high and remains high for the northern Gulf in spring when dissolved oxygen in bottom waters starts to decline. The ratio is above the Redfield ratio of 16:1, indicating the water has lots of nitrogen and not enough phosphorus, making phosphorus the limiting element, the one that sets the rate of algal growth.

The January 2004 draft of the 42-page report had strong language about the importance of phosphorus in controlling algal growth in the northern Gulf and about nitrogen not being the answer for solving the dead-zone problem. On page 37, it says, "There is a lack of compelling evidence that reduction of nitrogen would reduce the supply of organic matter fueling hypoxia"; then on the next page, "The CENR reports placed emphasis on levels of nitrogen reductions in the MRB and ARB [basins of the Mississippi and Atchafalaya Rivers, respectively], which cannot be supported by the available information." After much pushback from scientists outside of the EPA, the final draft of the report, now 25 pages, came out eight months later with more balanced language about the importance of both nitrogen and phosphorus.[7] It didn't explicitly criticize the CENR. But it still had lots about high nitrogen to phosphorus ratios and phosphorus limitation.

The report was immediately criticized by Don Boesch and two other leading scientists working on nutrients and the hypoxia problem. One was Don Scavia, who was involved in setting the 5000 square kilometer target for the dead-zone size (Chapter 5). The third, Bob Howarth of Cornell University, was quoted by a reporter as saying the phosphorus argument in the plan was "pretty naïve."[8] In their letter to the EPA, which Don Boesch shared with me, the three scientists pointed out flaws in how the report used nutrient ratios to argue for phosphorus limitation. One flaw is concentrating on just one form of phosphorus, phosphate, while ignoring other forms and sources of the element. Also, algae are incredibly proficient at scavenging phosphorus even when concentrations are very low. Other criticisms were discussed by Don in a letter published in *Science*,[9] one of the most prestigious journals in the discipline, and in a review of the topic published a couple years later.[10] Eventually, EPA decided to withdraw the report and not publish it.

But not before it made an impact. In addition to two pieces in *Science* and the review, the unpublished EPA report colored the recommendations from the Hypoxia Task Force. Recall that the Harmful Algal Bloom and Hypoxia Research and Control Act, first passed by the US Congress in 1998, set up a Task Force consisting of representatives from 12 states in the Mississippi River basin, five federal agencies, and the National Tribal Water Council.[11] The Task Force report in 2000 recommended that in order to reach the hypoxia area target of 5000 square kilometers, nitrate loading needs to be reduced by 30 percent. After the EPA report, and perhaps even more importantly, a change in the scientists making up the Hypoxia Advisory Panel, the 2008 Task Force report recommended reducing both nitrogen and phosphorus loading by 45 percent.

I think it's clear that while the EPA report went overboard about phosphorus, there is an argument for considering it even when the focus rightly remains on nitrogen. The topic was the focus of a seminar Katja Fennel recently gave at my university, and I caught up with her after to get her backstory.[12]

Katja's academic training started in mathematics at the University of Rostock, which at the time was in the German Democratic Republic, that is, East Germany. That explains why her English has a hint of a non-American accent, while her years in the United States and Canada explain why it's only a hint. Katja remembers having to sit through compulsory lectures about Marxism, even though the Berlin Wall fell early during her undergraduate years. She stayed at Rostock for graduate work, first in numerical mathematics and then for a PhD focused on a model of nutrients and plankton in the Baltic Sea. Although scientists at the nearby Leibniz Institute for Baltic Sea Research were working on hypoxia in the Baltic Sea, Katja didn't start thinking about the problem until she moved to Rutgers University, New Jersey, in 2002. Now she is a professor in the Department of Oceanography at Dalhousie University in Nova Scotia, Canada.

Katja Fennel has worked on modeling the effects of nutrient limitation on controlling the size and extent of hypoxia in the Gulf. She has used data from other investigators showing that the Mississippi River has high levels of nitrogen relative to phosphorus, as mentioned before. The ratio for two forms of these nutrients, nitrate and phosphate, are sometimes as high as 100:1, much higher than the Redfield ratio of 16:1. Those numbers indicate that there is relatively more nitrate than phosphate and that phosphorus limits algal growth, at least at first when river water flows into the Gulf. Then nitrogen limitation takes over further away from the coast. Work by Fennel and colleagues suggest that this temporary limitation by phosphorus may lessen hypoxia or at least change its location.[13] Because uptake of all nutrients is slowed down by phosphorus limitation, more nitrate is left over to boost algal growth downstream. As a result, the peak in algae is shifted further away from the mouths of the Mississippi and Atchafalaya. Working with Arnaud Laurent, Fennel built a mathematical model to explore the combined effects of nitrogen and phosphorus limitation on hypoxia in the Gulf.[14] The model does a good job in reproducing the area covered by hypoxic waters over the years. That gives some confidence that the model is right about how much nitrogen and phosphorus loading would have to be reduced to reach the 5000 square kilometer target. In their 2018 study, Fennel and Laurent say that nitrogen loading would have to be reduced by 63 percent and phosphorus by 48 percent. The needed reduction in nitrogen loading seems much greater than concluded previously by the Task Force and by other studies published in the scientific literature. But the uncertainty around the 63 percent estimate is rather high: plus or minus 18 percent. Fennel and Laurent say their estimates are "consistent" with previous estimates, and I think they're right. The most recent estimate is even higher.[15] In the end, it really doesn't matter.

I say that because nitrogen loading into the Gulf has not decreased much at all.[16] Even the low target of a 30 percent reduction in nitrogen loading agreed on twenty years ago has not been met.[17] That target was supposed to have been achieved in 2015. Realizing in 2013 that the target wasn't going to be reached, the Task Force moved the deadline to 2035 and set an interim target of reducing both phosphorus and nitrogen nutrients by 20 percent by 2025. The target area of 5000 square kilometers remains, and the approach taken by state and federal governments hasn't changed. The EPA and the states have resisted setting enforceable limits on nitrogen and phosphorus pollution. Even the state of Louisiana has been reluctant to call the northern Gulf "impaired" because it may trigger a regulatory process it doesn't want.

I suspect Don Boesch only grudgingly accepts a role for phosphorus, and he certainly wasn't happy about the EPA report. When he emailed the report to me, he mentioned the "painful memories." He felt the report "confused the Task Force process" and provided another "good excuse to waste a couple of years arguing about the science and demanding more proof and models before actions." He's probably right. But the cynic in me thinks that even without the fig leaves provided by the report and the scientific papers, agribusiness groups and others would have figured out other excuses to continue business as usual. We now have a consensus among scientists about the cause of the Gulf dead zone, yet too little is still being done to solve the problem.

"Shootout at the OK Corral"

The limiting-nutrient debate we just saw in the Gulf of Mexico was going on at about the same time in Sweden. There the debate wasn't about which experimental approach is best, because ratios and the bioassay experiments have given similar results in the Baltic.[18] It was focused squarely on which nutrient, phosphorus or nitrogen, limits algal growth in the Baltic. The answer would affect upgrades (or not) of wastewater-treatment plants in Sweden and regulations about fertilizer use and manure disposal. I mentioned before that the Baltic is limited by nitrogen, but there's more to the story. The early studies focused on phosphorus. A phosphorus-limited Baltic would fit with the paradigm that freshwaters are limited by phosphorus whereas marine waters are nitrogen-limited (Chapter 6); the salinity of many regions of the Baltic Sea is low, a third or less than typically found in the oceans. As we saw in Chapter 4, S. H. Fonselius in the late 1960s thought hypoxia in the Baltic was caused by excessive phosphate. Then, soon after nitrogen limitation was discovered in US coastal waters, a 1973 study found that inputs of deep water ("upwelling") into the surface layer of the Baltic lead to excessive phosphorus and inadequate nitrogen nutrients.[19]

So, what's the answer? Nitrogen or phosphorus? In spite of spending billions of Swedish krona (over $1 billion in today's money) on wastewater-treatment plants in the 1970s alone,[20] the country still had serious eutrophication problems along its coasts and a spreading dead zone in the Baltic Sea. The Swedish government wanted to know what to do next. In the spring of 2005, the Swedish EPA formed a committee of five outside experts to settle the debate and make some recommendations. The committee read the many published and unpublished papers on the topic, heard from Swedish researchers working on the Baltic, and did their own analyses of data to look for long-term trends.[21] Page seven of the committee's report has a surprising statement about the tone of the discussions. I wasn't surprised to learn they were "lively." The surprise, at least to read in a scientific report, was that the debate was "at times acrimonious." One of the five outside experts would later call the nitrogen versus phosphorus debate "the shootout at the OK Corral."[22]

The gunslingers on the committee came to Stockholm armed with their answers about which nutrient was most important. On the phosphorus side were two limnologists, Robert Hecky and David Schindler. American-born Hecky was working in Canada at the time, but is now at the University of Minnesota in Duluth on the shores of Lake Superior. Although Schindler is also originally from the United States, he has made his career working at the Canada's Experimental Lakes Area (Chapter 6). Studies at those lakes were key in showing phosphorus limitation in freshwaters. Armed to the teeth with those studies and others, Schindler was adamant and forceful in arguing for phosphorus being the answer for the Baltic Sea. He was the one who compared the debate to a shootout and published his argument soon after the Stockholm meeting.[23] On the nitrogen side were Don Boesch, who we just saw in the Gulf of Mexico, and Sybil Seitzinger, who at the time was at Rutgers University in New Jersey. The fifth member and chair of the committee, Charles O'Melia, was at Johns Hopkins University until he passed away in 2010. As an environmental engineer who worked on water treatment, unaligned with the limnologists (Schindler and Hecky) or the oceanographers (Boesch and Seitzinger), mild-mannered O'Melia was neutral in the fight. He must be the person Schindler's book says was "on the fence," perhaps a safe place in a corral to watch a gunfight.

The report pointed out that the discussion got testy at times because of "different results from different regions, different interpretations of results and the relative roles of the above factors [nitrogen and phosphorus nutrients], and a shortage of definitive data." We now have more data to answer some of the questions. We now know that different regions of the Baltic are limited by different nutrients. The low-salinity water of the Bothnian Bay is phosphorus-limited, as is the Bothnian Sea in spring.[24] However, the Bothnian Sea, which is the northern lobe of the Baltic, is limited by nitrogen in the summer. Likewise for the main sections of the Baltic, the Baltic Proper, most of the year. In the Baltic Proper, low nitrogen and warm temperatures, however, lead to cyanobacterial blooms in the summer, touching on the most contentious part of the report. One specific question put to the committee by the Swedish EPA was whether reducing nitrogen loading would cause more cyanobacterial blooms and more hypoxia. Blooms of the wrong cyanobacteria can be even nastier than algal blooms, and they can worsen water quality by producing foul-tasting organics and toxins. The fear was that reducing nitrogen nutrients would select for nitrogen-fixing cyanobacteria, which have an advantage over algae unable to make usable nitrogen from nitrogen gas. A prominent Swedish oceanographer even suggested that nitrate should be added to the Baltic in order to control the cyanobacteria.[25]

In fact, nitrogen was purposely added to one part of the Baltic in the early 2000s as part of an incredible "experiment" in Himmerfjärden, a fjord-like estuary about 60 kilometers south of Stockholm. Himmerfjärden starts with

Lake Mälaren in the north and ends in the Baltic Sea to the south. When water levels are high in the lake, fresh water flows into Himmerfjärden via a sluiceway, making the fjord's water brackish, with a salinity of 0.6 percent, much lower than the 3.5 percent in full-strength seawater. Beginning in 1974, wastewater that previously went into Lake Mälaren was diverted to Himmerfjärden, but at least it was treated to remove nearly all of the phosphorus. An effective method for removing nitrogen wasn't implemented until 1998, successfully reducing inputs to 150 tons per year from a high of 900 tons per year. Around this time, the nitrogen-fixing cyanobacterium *Aphanizomenon* took off and became abundant in the summer, according to data from Ragnar Elmgren and Ulf Larsson, two ecologists now retired from Stockholm University. The "experiment" was to add nitrogen, not just to a bottle or a big tank or even a lake, but to an entire estuary.

In 2001 and 2002, Elmgren and Larsson asked the wastewater-treatment plant to discharge nitrogen into Himmerfjärden to see how it would affect the cyanobacterium and the algae. Sure enough, with the nitrogen addition, the cyanobacterium declined. The experiment was repeated in 2007 and 2008 with similar results.[26] It sounds like the freshwater guys were right: concentrate on stopping phosphorus pollution and forget about nitrogen. Page seven of the Stockholm report put it more bluntly: "controlling nitrogen is a waste of money and the focus of reductions should be on phosphorus alone." However, cyanobacteria aren't the complete story. Although it's true that the cyanobacterium *Aphanizomenon* did become more abundant with lower nitrogen inputs, the algae declined dramatically. Algae followed the nitrogen and increased in abundance during the 1990s when nitrogen inputs were high and then decreased when nitrogen removal became effective in 1998. The useable nitrogen made by the cyanobacterium was small compared to other nitrogen sources and wasn't enough to fuel growth of the algae. Although a cyanobacterial bloom is generally not good, *Aphanizomenon* is not toxic and doesn't cause the water-quality problems seen with other cyanobacteria. The freshwater gunslingers were right about the cyanobacteria, yet their extreme position of not doing anything about nitrogen was not. Less nitrogen did result in more cyanobacteria, but it also ameliorated the eutrophication problems caused by the algae.

The reason why both nitrogen and phosphorus nutrients must be controlled is illustrated by a simple model, the "vicious circle" (Fig. 7.2), that combines relevant facets of microbiology and aquatic chemistry.[27] Inputs of nitrogen and phosphorus nutrients stimulate the growth of algae and

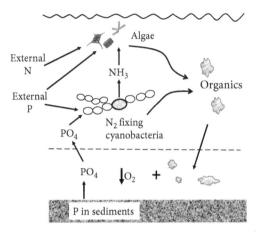

Figure 7.2 The vicious circle. The input of external nitrogen (N) and phosphorus (P) stimulates the growth of algae and nitrogen-fixing cyanobacteria. The organic material they make uses up oxygen ($\downarrow O_2$) and promotes the release of phosphate (PO_4) from sediments, which stimulates more growth and more hypoxia.

cyanobacteria, which produce the organic material that fuels the depletion of dissolved oxygen from the bottom. As the sediments lose oxygen and turn anoxic, they release phosphate that was previously complexed with iron when dissolved oxygen was plentiful. The released phosphate then stimulates growth of the nitrogen-fixing cyanobacteria, which produce more organic material, starting the vicious circle over again. To keep the circle running, inputs of external nitrogen are needed to make up for the nitrogen lost as nitrogen gas during denitrification, which kicks in as dissolved oxygen disappears. So, stopping the inputs of nitrogen from wastewater-treatment plants and other sources is key to solving the problems caused by eutrophication.

The OK-Corral fight in Stockholm ended in a draw, judging from the report, and both sides went back home convinced they were right. Yet Sweden continued to install tertiary wastewater-treatment plants that remove nitrogen nutrients as well as phosphorus, but nitrogen removal has been less effective. Although treatment plants released over 25,000 tons of nitrogen in the early 1980s, in 2016, the release was lower by about 50 percent—impressive, but the reduction was less than achieved for phosphorus.[28] After maxing out at about 7000 tons released each year in the late 1970s, phosphorus released by the treatment plants has decreased by over 95 percent.

Wastewater-treatment plants have been a bright spot in the fight against eutrophication and hypoxia in Swedish coastal waters and around the world.

Before leaving the Baltic, it is useful to compare it to the Gulf of Mexico. Although the two dead zones share several similarities, one difference is the nitrogen-fixing cyanobacteria in the Baltic. Relatives of these microbes are in the Gulf, but not the variety that fix nitrogen using heterocysts, which are cells specialized for carrying out nitrogen fixation. The heterocyst-bearing cyanobacteria may be selected against by the Gulf's high salinity. The two dead zones also differ in how much of the surface waters is limited by phosphorus. That element limits algal growth and affects hypoxia in the Gulf only where the nitrogen-laden Mississippi and Atchafalaya Rivers debouch. By contrast, the Baltic Sea has more low-salinity water, and phosphorus limitation is more common. Even in the Baltic Sea, however, nitrogen becomes limiting further offshore.

Phosphorus is even more important in the next type of waterbody to be discussed, lakes.

A Dead Lake

I suppose a book about the dead-zone problem could focus on marine waters and leave out lakes. The areas covered by hypoxia in marine regions like the Gulf of Mexico and the Baltic Sea are huge, with equally large environmental consequences. They grab headlines worldwide. But even if the focus is on marine systems, lakes provide another argument for controlling fertilizer runoff and leaching of nutrients from manure. Those nutrients should be stopped not only to reduce a dead zone in marine waters hundreds of kilometers and miles away but also to solve eutrophication problems in nearby rivers, reservoirs, and lakes. I also want to talk about lakes because I like them. I grew up swimming and fishing on small lakes in Wisconsin like Shawano Lake, which is less than an hour drive northwest of Green Bay. When I was a kid, Lake Michigan was my ocean. Given my lack of seaworthiness, maybe I should have been a limnologist. This is the place to discuss lakes because the size of their dead zones is set by phosphorus.

Edward Asahel Birge, a pioneer limnologist working at the University of Wisconsin in Madison, was one of the first to look at oxygen in lakes. First hired in 1875 to be an instructor in natural history, Birge was appointed dean in 1891 and acting president in 1903. After being passed over for the permanent presidential job, he found time in 1906 to give a presentation about Lake Mendota at the thirty-fifth annual meeting of the American Fisheries Society.[29] He began by saying that the "conditions of life" weren't known well enough to understand the "lower animals" in Wisconsin lakes. So, he set out to explore those conditions. The rest of his presentation is what could be heard today in Limnology 101 in the university's Birge Hall. Birge describes how temperature and dissolved oxygen were the same from top to the bottom of Lake Mendota on April 22, 1905, when the world was astounded by the Wright brothers' heavier-than-air machine flying over 20 miles. Then by August 31 of that year, the surface layer had warmed to 23 degrees Celsius, while the bottom layer remained a cool 12 degrees (Fig. 7.3). It had become stratified, with only the temperature difference causing the gradient in density, without any help from the salt found in marine waters. Birge says the bottom stratum, what limnologists call the hypolimnion, was "cut off from the air by the upper stratum," with a predictable effect on dissolved oxygen levels at the bottom. As organic debris raining from the surface is degraded, oxygen disappears from the hypolimnion. A few years later, Birge and his long-time collaborator, Chancey Juday, reported that in summer, the hypolimnion of many lakes in Wisconsin (they looked at 156

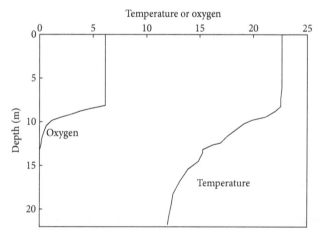

Figure 7.3 Temperature (degrees Celsius) and dissolved oxygen (milligrams per liter) in Lake Mendota on August 31, 1905, as measured by E. A. Birge. The lake was stratified because the surface water was much warmer and less dense than the cold bottom layer, preventing the surface water from mixing with the bottom water. As a result, the bottom water lost all of its dissolved oxygen.

lakes but not Shawano), a lake and pond in Massachusetts, and a small lake near Nantua, France, had little to no dissolved oxygen.[30]

Some of the oxygen loss from lakes is natural, some of it isn't. The bottom layer of lakes can go hypoxic or anoxic with natural inputs of organic material from the surrounding land. Nutrients like nitrate and phosphate from the lake's watershed can also fuel algal growth in surface water and oxygen depletion in lake bottom waters. But people also potentially add more organics and hasten the inflow of nutrients. A study used the presence or absence of lamination in sediments, which we saw applied in the Baltic Sea, to look at the occurrence of hypoxia in 365 lakes worldwide from the year 1700 to today.[31] The scientists found that the number of lakes with hypoxic hypolimnions was the same (about 290) for hundreds of years. Presumably this hypoxia is due to natural causes. Then the bottom waters of more and more lakes started to turn hypoxic around 1850 until eventually now an additional 71 lakes have hypoxia or anoxia. The upswing in hypoxia couldn't be explained by changes in temperature or precipitation. Rather, the scientists found a very tight correlation between the rise in lacustrine hypoxia and the release of phosphorus from fertilizer and other sources. From 1800 to 2000, the total release of phosphorus and fertilizer phosphorus increased by a factor of five and seven, respectively. Hypoxia became more common because the lakes received more phosphorus.

Phosphorus pollution and its consequences are vividly on display in Lake Erie, the shallowest and smallest in volume of the five Great Lakes. The lake was declared dead in the 1960s and 1970s, even though life in the form of algae was so exuberant that Erie water was said to be "too thick to pour but too thin to plow."[32] The lake had turned into pea soup because of the phosphorus from inadequately treated wastewater from a human population in the lake's basin that had tripled over 50 years. The nutrient also came from agriculture. Use of phosphorus fertilizer in one state on Lake Erie's shore, Ohio, increased threefold from 1930 to 1979.[33] As a result, between 1948 and 1962, phosphate levels in the lake increased by nearly five times, much more than the 30 percent jump in nitrogen levels. With those nutrients came higher algal growth. The western and central basins of the lake were so coated with cyanobacteria in the summer that the surface looked as if green paint had been dumped on it. To others it seemed the color was the lime green of the wicked witch's face in the "Wizard of Oz." The pollution along with overfishing killed the sport and commercial fisheries of the lake. Fish prized by anglers like blue pike, cisco, and whitefish declined while rougher species like sheepshead, alewife,

and carp took their place. The algal blooms also led to hypoxia in bottom waters. As with other hypoxic waters, Lake Erie was far from dead, but the life it
harbored could be appreciated only by a microbiologist.

Lake Erie has come back to life, although it is still not in perfect health
today. The lake's recovery started in 1972 with the phosphorus abatement
program set up by the Great Lakes Water Quality Agreement between
Canada and the United States.[34] That year also saw the passage of the Clean
Water Act in the United States, which funded more and better wastewater-
treatment plants. Phosphorus-rich detergents were restricted, although
complete bans for most US states and Canada came much later.[35] Those
restrictions led to a drop in phosphorus inputs into Lake Erie even below
the Agreement's target of 11,000 metric tons per year. As phosphorus levels
declined, so too did the algae, and dissolved oxygen returned to bottom waters. However, the lake's recovery stalled in the mid-1990s when inputs of a
form of phosphorus started to inch back up, and the algae and cyanobacteria
returned. Today, most of the phosphorus is not coming from "point sources"
like treatment plants, places you could point a finger at or mark with a dot on
a map, but rather "nonpoint sources" like agriculture. In the western basin,
the shallowest part of Lake Erie where algal blooms have always been the
worst, the point sources now account for only 9 percent of total phosphorus
inputs, whereas nonpoint sources contribute about 90 percent.[36]

The nonpoint sources empty into Lake Erie's western basin via the Maumee
River. The watershed of the Maumee was once the Great Black Swamp,
which stretched over 160 kilometers, covering about 4000 square kilometers
with forests, grasslands, and wetlands in northwestern Ohio and eastern
Indiana.[37] The swamp was colored black by the rich muck that had accumulated over millennia in poorly drained areas. Although lands to the north,
south, and east were being settled, the Swamp was considered uninhabit-
able and impenetrable by Europeans. A missionary traveling to Detroit in
October 1791 wrote of the "deep swamps and troublesome marshes" and the
many miles "where no bit of dry land was to be seen, and the horses at every
step wading in the marsh up to their knees."[38] Malaria was endemic. While
most viewed the Swamp as "land not worth a farthing," others could see its
potential. Settlers eventually came to fell the sycamores and elms, providing
wood for corduroy roads and houses, and fuel and ties for Ohio's railways. To
dry out the land, farmers dug ditches alongside fields and channels under-
neath the fields to carry away excess water. The first drainage tiles lining the
channels were ceramic, but then they were made with the clay found beneath

the Swamp's soil. By 1880, northwest Ohio had more than 50 drainage tile factories. As settlers continued to cut trees down and dig ditches and lay tiles to drain the land, over five short decades the Swamp was transformed into cropland and pasture.

By 1920, the Great Black Swamp was no more. It was replaced by farms growing mostly corn and soybean. Instead of the Swamp sucking up nutrients from the Maumee River, the region now is a source of those nutrients, which come from the fertilizer used to grow the corn and soybean and from the manure excreted by the many cattle, dairy cows, hogs, and chickens raised in the region. These nutrients are the reason why Lake Erie still has hypoxia and algal blooms that reoccur each summer. Cleveland, Ohio, which is on the shores of the lake's central basin, keeps an eye on the Erie dead zone because it affects the quality of the city's drinking water. The nasty tasting chemicals that form in hypoxic water have to be removed by activated carbon or other expensive treatments.[39] But Toledo, Ohio, has had an even more serious problem with its drinking water.

The problem is caused by a type of bloom-forming cyanobacteria, *Microcystis*, that hasn't been abundant in Lake Erie until recently. The microbe shares many characteristics of the cyanobacteria previously plaguing the lake, most notably having a shape that deters herbivores, making it a poor base for the lake's food chains. Rather than occurring as filaments, the configuration taken by other bloom-forming cyanobacteria, *Microcystis* clumps together into colonies big enough to be visible to the naked eye, held together by a thick, inedible mucilage. But the reason there is so much interest in *Microcystis* is that it produces a chemical, microcystin, which is toxic to invertebrates, fish, and animals, including us. Exposure to microcystin-contaminated water has been linked to fever, abdominal pain, vomiting, and diarrhea; one of the more than 70 versions of the toxin, microcystin-LR, is also thought to be a carcinogen.[40] Its chemical structure makes it difficult for bacteria to degrade, and even boiling doesn't destroy it.

Now you can understand why the city of Toledo, located at the mouth of the Maumee River, is concerned when *Microcystis* blooms in Lake Erie. It's the source of the city's drinking water. Summertime blooms of *Microcystis* had become more frequent in the western basin since the mid-1990s.[41] To better quantify these blooms, the US National Centers for Coastal Ocean Science has used satellite images to produce a bloom severity index.[42] The index was usually low, 1.6 on average (10 is the worst), for 2002–2007, but for the last ten years, the average has been over six. It was 6.6 in 2014. On

August 2 of that year, over 500,000 residents of the Toledo metropolitan area woke up to find they couldn't use the water coming out of their faucets;[43] they weren't supposed to drink it, brush their teeth, or cook with it, or give it to their dog. Water leaving the city's treatment facility was found to have dangerously high levels of microcystin. Two days later, researchers from the Great Lakes Environmental Research Laboratory sailed out to the intake site for Toledo's water in order to figure out what was causing the contamination. They found *Microcystis* and high levels of microcystin, including the carcinogenic version, microcystin-LR. Although Toledo residents couldn't use their water for only a couple days, the damages from the restrictions were estimated to be at least $65 million.

It is not entirely clear why *Microcystis* blooms have become more common over the last few decades, but one answer may be nitrogen nutrients. It's pretty clear the answer is not total phosphorus, which had reached a steady level, more or less, in the years before *Microcystis* took over in Lake Erie. The surge of *Microcystis* has been tied to the higher inputs of one form of phosphorus, soluble reactive phosphorus. This form is called "reactive" because it is readily used by cyanobacteria and algae. However, other work has suggested that *Microcystis* does better than its cyanobacterial competitors when soluble reactive phosphorus is low. Other possible answers for *Microcystis* infestations include warming temperatures and the invasion of the Great Lakes by zebra and quagga mussels.

I think part of the answer is organic nitrogen. Unlike the cyanobacteria previously dominating blooms in the lake, *Microcystis* is not able to get its nitrogen via nitrogen fixation and instead uses a variety of fixed nitrogen chemicals in the lake water. So, inputs of those chemicals, along with less soluble reactive phosphorus, would favor *Microcystis* over other, nitrogen-fixing cyanobacteria. The input of total fixed nitrogen into Lake Erie hasn't changed much over the last two decades, but one form of fixed nitrogen, organic nitrogen, has increased.[44] One organic nitrogen chemical is urea, commonly used in fertilizers by farmers in Ohio. Inorganic nitrogen may also have a role. Levels of microcystin in Lake Erie and other lakes correlate, not with phosphate, but with inorganic nitrogen; laboratory studies have shown that *Microcystis* produces more of the toxin when given inorganic nitrogen.[45] Whether it's organic or inorganic, the message seems clear that nitrogen can't be ignored if Toledo is to have safe drinking water. Although the latest U.S. Action Plan for Lake Erie continues to focus on controlling phosphorus inputs, it recognizes the connection between nitrogen and toxicity of the cyanobacterial blooms.[46]

Toledo residents can be excused for caring more about the safety of the water coming out of their faucets than about the level of dissolved oxygen in Lake Erie. But the two are connected. Both are caused by exuberant cyanobacteria stimulated by excessive nutrients. In addition to toxins, these organisms produce organic materials that sink and deplete oxygen in bottom waters. Cyanobacterial blooms can lead to hypoxia, and then hypoxia feeds the bloom by releasing phosphate. The same chemistry that releases phosphorus from anoxic sediments of the Baltic Sea works in Lake Erie. Although I haven't seen "vicious circle" mentioned in publications about hypoxia in waterbodies other than the Baltic, the phrase is applicable to Erie and other lakes. More hypoxia means more phosphorus that feeds more cyanobacteria and algae, which turn into more hypoxia. Solving the lake's dead-zone problem would ensure that Toledo doesn't see another summer of 2014.

I almost forgot about the evil ecologist and his question about which type of fertilizer he should use to start a dead zone in coastal waters. Even after reading the studies about the importance of both nitrogen and phosphorus, he decides that nitrogen fertilizer is the way to go. Marine waters are mostly limited by nitrogen, and besides, he could get a better price if he bought lots of only one type of fertilizer. Let's not tell him that the most effective strategy depends on the salinity of his targeted water body. A nitrogen-first approach does seem best for salty waters like the Gulf of Mexico, but we saw that phosphorus shouldn't be ignored. At the other end of the spectrum, algae in lakes are limited mainly by phosphorus, as is the extent of hypoxia. The basins and bays of the Baltic Sea vary in salinity and whether nitrogen or phosphorus is the limiting element. Although we can argue about which nutrient is the limiting one, we can't just concentrate on one and ignore the other if we want to solve the dead-zone problem.

8

Fish and Fisheries

As scientists worked at figuring out what causes dead zones, others were exploring the effects of hypoxia on aquatic organisms and habitats. The most obvious had been long known. In his "History of Animals," Aristotle observed that fish die of suffocation when kept without enough water just as animals suffocate when they don't have enough air. In the early 1800s, Humphry Davy, famous for his work on nitrous oxide, including experiments on himself, discovered that oxygen was the dissolved gas that keeps fish and other aquatic animals alive.[1] More recently, in laboratory experiments, we have learned that aquatic animals vary greatly in their sensitivity to dissolved oxygen deprivation, as I've pointed out before—ranging from burrowing shrimp, which can survive if only briefly in dead-zone levels of oxygen, to cod, which are stressed when oxygen slips even a bit below 100 percent saturation.[2] But it's harder to see this range of hypoxia effects with entire populations of animals in the real world. Scientists can't do the same type of controlled experiment with a bay or coastal ocean as they can with a fish or crab in a bottle. The artificial aeration of a waterbody, which has been done a few times (see Chapter 10), comes close to the experiment I have in mind; but these aeration projects so far have been focused on solving a practical problem, not on science. Even if the projects had focused on science, it would be difficult to extrapolate from artificial aeration of a small pond or inlet to dead zones like the size of those in the Gulf of Mexico or the Baltic Sea.

Of course we know the most extreme result of hypoxia—death—does occur in the real world, not just in the laboratory. There have been an increasing number of fish kills around the world, from the subarctic (the Mariager Fjord in Norway) to the subtropical (Richmond River Estuary in New South Wales).[3] In Chapter 2, I mentioned the death of clams and cockles in the Baltic Sea and the dead goby and flounder stinking up beaches when the northern Black Sea turned hypoxic. More recently, about 1.5 million sardines died after being trapped in hypoxic (and acidic) waters of King Harbor in Redondo Beach, California,[4] and more than 25,000 carp turned up dead in the St. Lawrence River.[5] Between 1980 and 1989, there were over 3,600 fish-kill events in US coastal states.[6] In US waters, many dead-zone victims, at least the most noticeable ones, are menhaden, which travel in large schools and conveniently float to the surface when dead, ensuring their deaths do not go unnoticed. Of the over 380 million fish killed by hypoxia in coastal waters of Texas from 1951 to 2006, over 70 percent were menhaden.[7] Some of the hypoxia-caused deaths for this fish may have been self-inflicted. Striped bass, bluefish, and other predators can herd large schools of menhaden into small, shallow embayments where the fish's own respiration strips the water of oxygen.

Fish and invertebrates trying to escape hypoxia can be an unexpected bonanza for people waiting on shore. In Mobile Bay, Alabama, on a hot August night, fish, shrimp, and crabs crowd into shallow waters where they are easily caught by gigs and nets or just scooped up by washtubs. It's a festive party, a Jubilee. Although known to Native Americans and described by Mobile's *The Daily Register* in 1867, the Jubilee wasn't fully explained until 1973.[8] When the tide and wind are right, the temperature high, and the sun down, oxygen disappears from deep waters in Mobile Bay, forcing fish and invertebrates to seek oxygen in shallows. Something similar happens off Namibia and the west coast of South Africa.[9] There it's called a lobster walk out. Thousands of rock lobster (*Jasus lalandii*) perambulate toward shore in search of oxygen, only to be stranded dry when the tide recedes. For the lobsters of Namibia and the fish, shrimp, and crabs of Mobile Bay, escaping the dead zone and death by suffocation is worth the risk of ending up in a bisque or jambalaya.

There is a common feature of the locations where death by hypoxia becomes noticeable to even a casual observer onshore. All of the locations are ponds, rivers, estuaries, or small bays: confined habitats without easy access to open, well-oxygenated water. The hemmed-in animals cannot escape. Most of the 3,600 fish kill events mentioned before fit that description: the top ten fish kills include a canal and a creek in Galveston, Texas; the Nanticoke River in Maryland; and the Little River in my state of Delaware. Mobile Bay with its Jubilee is another confined habitat. Another example is fish kills in aquaculture facilities in the Philippines and elsewhere,[10] a problem exacerbated by nutrients released by these facilities. In China, nutrient pollution by aquaculture may rival pollution by livestock production on land.[11] Regardless of the source of nutrients, fish in pens and cages have no chance of escaping hypoxic waters.

What about the second-biggest dead zone in the world, the Gulf of Mexico? I mentioned the fish kills in "Texas coastal waters," which sounds like part of the Gulf. In fact, the fish kills were in Galveston Bay, Matagorda Bay, San Antonio Bay, Copano and Aransas Bays, Corpus Christi Bay, and Laguna Madre—all bays except the last, a shallow lagoon with only one channel connecting it to the Gulf. Fish were killed by hypoxia in those bays, but not by the Gulf dead zone. The *Houma Daily Courier*, the southern Louisiana newspaper that first used "dead zone" to describe hypoxic waters, has had several front-page stories over the years about fish kills, but those have been in bayous or in lakes, not in the Gulf. There have been fish kills near Grand Isle when hypoxic waters trap fish near the beach.[12] But what about offshore in the Gulf?

Irony in the Gulf

The Gulf dead zone overlaps with the "Fertile Fisheries Crescent," an arc of coastal waters stretching from Texas to Florida, with Louisiana in the middle, one of the most productive fishing grounds in the world. There, fishermen can catch grouper, five kinds of snapper, king mackerel and blue marlin, menhaden, yellowfin tuna and blackfin tuna, redfish, cobia, wahoo, amberjack, sharks, and speckled trout. The Gulf has over 1440 species of finfish, making it one of the most diverse habitats in the United States.[13] Along with those fish are blue crabs, brown shrimp, and white shrimp, all worth about $700 million dollars in 2017.[14] Although there are no seals, sea lions, or sea otters, the Gulf has bottlenose dolphins and the Bryde's whale (*Balaenoptera edeni*), as well as Kemp's ridley, loggerheads, and three other species of turtles.[15] Onshore ornithologists can count on seeing the reddish egret, brown pelican, and three different species of plovers, while offshore are the common tern, the common loon, the least bittern, and more—nearly 400 bird species in total.[16] Ironically, right in the middle of all this life is a dead zone.

Hypoxia and all those animals are there for the same reason: the nutrients brought down to the Gulf by the Mississippi River. Those nutrients fuel algal growth that eventually feeds all the animals in the Gulf or living along its shores. Just as fertilizers enhance crop production on land, Mississippi River nutrients enhance algal production in the sea. More algae mean more redfish, amberjack, brown shrimp, and just about everything else up the food chain. In general, there is a positive relationship between nutrient inputs and fish landings. But at some point, there is too much algal growth. With excessive nutrients come excessive algae that cannot be eaten by the organisms desired by anglers or nature lovers. As we've seen already, the organic material from those algae makes its way to bottom water where it's degraded by bacteria while using up the dissolved oxygen. As a result, the Gulf has both the Fertile Fisheries Crescent and a dead zone.

So, you'd think hypoxia would have a big effect on fish and fisheries in the Gulf. There are many well-documented effects of hypoxia on fish, including altered foraging, reduced growth rate, impaired reproduction, higher mortality, and changes in predator-avoidance behavior.[17] Those effects must be seen by anglers and commercial fisheries in the Gulf, right? One answer is from a captain of a charter fishing boat, Tommy Pellegrin:

> Every year, they put out a news release about the Gulf of Mexico dead zone. I've been running charters in the Gulf for decades. In all that time, I've never returned home without catching fish. When I'm fishing offshore, I probably come back with an average of 300 to 400 pounds of fish. Sometimes, we catch more than 1,000 pounds.[18]

That anecdote is not too far off from the science. Certainly, most sessile invertebrates are killed by hypoxia in the Gulf, especially the larger, long-lived organisms that burrow into sediments.[19] These are replaced by smaller, short-lived worms (polychaetes) able to survive short bouts of hypoxia or to come back quickly when dissolved oxygen returns. Overall diversity of benthic invertebrates has decreased as dead zones expanded and became more common in the Gulf over the last half century.[20] But the impacts on the fish and shrimp favored by Captain Pellegrin and targeted by fisheries are harder to see.

Except when raw sewage is involved, hypoxia effects on fisheries landings are less clear-cut than expected from the many studies showing the death and destruction caused by low oxygen.[21] Fisheries landings of mobile species tend to be high in nutrient-rich waters like the northern Gulf of Mexico and East China Sea in spite of hypoxia (Fig. 8.1). That conclusion is from an influential review by Denise Breitburg and colleagues published over ten years ago, but it remains true today, albeit with some important qualifications and additions. We will soon see why it is difficult to show dead-zone effects on fisheries, and we'll go over evidence of some real impacts. It's not a straightforward story.

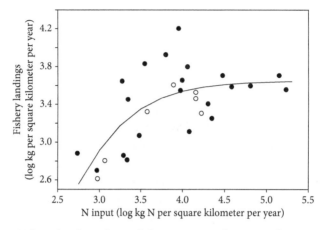

Figure 8.1 Fishery landings for mobile species as a function of nitrogen (N) inputs for several regions around the world, including the Gulf of Mexico and Baltic Sea. It is difficult to see any difference in landings in regions with a dead zone (open circles, greater than 39 percent of the region's area covered by hypoxic bottom water) from other water bodies with normal oxygen levels (filled circles).

The lack of a simple, clear-cut cause and effect has hindered efforts to marshal support for reducing the dead-zone area in the Gulf and elsewhere. Why should we care about hypoxia if it doesn't affect the fish we catch and the seafood we eat? That's the argument used by a fishery scientist one floor below where Nancy Rabalais, aka "Queen of the Dead Zone," has her office at Louisiana State University.[22] Not a good sign if Nancy can't convince him. Don Scavia, who has done a lot of work on hypoxia and helped to shape legislation in the United States to address the problem, thinks the lack of a clear effect is the "Achilles heel" of the fight against dead zones, at least the one in the Gulf.[23] The Gulf isn't the 19th-century River Thames where cholera and the river's stench were powerful reasons to fix the dead zone. The dead-zone effects in the Gulf are more complicated than a waterborne disease or sewage's stink, and that's a problem. An argument that could be distilled to a single tweet or a bumper sticker (maybe "Dead Zones Kill Fish and Fishery Jobs," written over a skull and cross bones) would be more persuasive and more easily grasped than one requiring pages of dense text, filled with equations and scientific jargon. So, without that bumper-sticker-sized message, we have to turn to a longer discussion of how hypoxia affects fisheries. At least there will be no equations.

The Bad—and the Good—for Fish and Fisheries

Ecologists face a big problem in trying to separate a hypoxia effect from the effects of the many other external forces shaping fish populations. One such force is recreational and commercial fishing. Of course, fishing has a huge impact on fish populations, evident from the decline of fish stocks overexploited by fishermen and the return back to natural levels when regulations and fishing limitations are imposed. The interaction between overfishing and hypoxia is perhaps best understood for Baltic cod. Because of good environmental conditions and low fishing pressure, numbers of the fish exploded in the 1980s, only to be decimated again by the end of the decade when intense fishing resumed.[24] New fishing regulations put into place in 2007 have contributed to the resurgence of the fish today, but Baltic cod are now leaner and growing more slowly than in the past, perhaps because of the proliferation of hypoxic bottom waters. Even in the Baltic, however, it has been difficult to separate out overfishing impacts from dead-zone effects. On top of fishing, natural mortality, vagaries in weather, and the randomness of nature complicate detecting a dead-zone effect.

Another complication is the movement of fish and other mobile species from hypoxia to well-oxygenated waters. Although escape is not possible in some bays and estuaries, it is in the Gulf of Mexico. Once in waters with adequate oxygen, the hypoxia-escaping species may even enjoy the benefits of the algal growth stimulated by eutrophication. In the Gulf, bottom-hugging fish and shrimp are tolerant of low-oxygen waters and thus don't need to go far from the dead zone, only a few kilometers, depending on the species.[25] They may want to stay close to waters with high biological production and rich benthic communities, the source of their food. Staying too close to the dead zone comes with costs, however. Atlantic croaker, for example, will not flee waters with only a bit more oxygen than the dead-zone level, but the fish's reproduction is impaired.[26] Costs or no costs, many factors cloud the relationship between oxygen levels and the apparent health of fish and fisheries.

Still another complication is that fishermen indirectly respond to hypoxia by following fish, crabs, and shrimp as they flee the dead zone. In the Chesapeake Bay, crabbers follow blue crabs as they move into waters with sufficient oxygen. In the Gulf of Mexico, the commercial purse seine fleet follows menhaden inshore and to the west as dissolved oxygen declines.[27] There is a similar but more complicated story with shrimp. The response of shrimp and shrimpers to hypoxia has been closely scrutinized because the fishery, especially for brown shrimp (*Farfantepenaeus aztecus*), is the one of the most valuable in the United States, pulling in over $500 million each year.[28] Like the crabbers following the crabs in the Chesapeake, shrimpers follow the shrimp in the Gulf. Nancy Rabalais showed me an incredible map of shrimp trawlers perched along the edge of the Gulf dead zone.[29] It's a kind of an open water Jubilee, the shrimpers scooping up shrimp as they escape hypoxic waters. Because shrimp are more concentrated at the dead-zone edge, their "catchability" is higher, and more shrimp are taken by the shrimpers.

So, hypoxia has a mixed effect on shrimp fisheries. Shrimp catch may or may not decrease as the dead zone expands, depending on the study and region.[30] Even though shrimp growth is slower in low-oxygen waters, that negative may be outweighed by the higher catchability of shrimp, making it appear as if hypoxia helps the shrimp fishery. Higher catchability may also attract more fishing effort, drawing in shrimpers from well-oxygenated waters, resulting in more shrimp being caught as hypoxia expands. It illustrates how hypoxia can cast a shadow on fishing efforts far from the dead zone, an impact easily overlooked or misinterpreted. For shrimpers, traveling further has its costs and brings in still another factor complicating the relationship between hypoxia and shrimp harvests. As one shrimper put it, "We have to travel further depending on how bad the dead zone is. . . . Forget about running home for a night or two because the expense is so much to go a great distance like that to catch shrimp."[31] Higher fuel costs can lead to lower fishing effort and lower shrimp harvests and higher prices for some sizes of shrimp.[32]

The intertwining of these factors and hypoxia in the Gulf of Mexico was explored by an influential study led by Marty Smith at Duke University in 2017.[33] Smith and colleagues tried to capture in a single economic analysis the effects of hypoxia on shrimp growth and mortality along with factors such as catchability, fuel prices, and the number of fishing trips for over 3500 vessels. To make a long story short, they didn't see an impact of hypoxia on shrimp landings in their data because the confounding effects of hypoxia on growth and mortality (negative) were balanced by higher catchability and greater fishing effort (positive). But the environmental economists did see that as the Gulf dead zone expanded and contracted over the years, so too did the price of large shrimp. As the area covered by hypoxia increased, large shrimp got more expensive. The authors calculate that a 1000 square kilometer expansion of the dead zone, which is about 6 percent greater than the average, translates into a 1 percent rise in the large shrimp price. Following the law of supply and demand, these shrimp are more expensive as they become rarer when hypoxia interferes with migration and slows down growth. Here finally is a study showing the negative impact of hypoxia on an important fishery in the Gulf.

There is another mechanism linking hypoxia with fish and fisheries. Remember how the dead zone in the River Thames and the Delaware River prevented fish like salmon and shad from migrating to essential spawning grounds? Even if the adult fish can stay in well-oxygenated waters, the species is doomed if it can't migrate to appropriate areas to spawn and reproduce. The same applies to the many species that use estuaries as nurseries. Unfortunately, the estuaries connected to the Gulf of Mexico aren't good examples. Shrimp use estuaries as a nursery, but the invertebrates leave them before hypoxia becomes a problem. After spawning in the Gulf, newly hatched shrimp enter estuaries of Gulf states in April and May when dissolved oxygen is sufficient. After growing to subadults, shrimp migrate back into the Gulf in May through August. This migration to deep waters can be blocked by the dead zone, and adult shrimp have to deal with hypoxia in those waters, but juvenile stages face fewer oxygen problems in their estuarine nurseries.

A better example of how estuarine hypoxia affects fish and a fishery is the Elkhorn Slough, a highly eutrophic estuary in central California.[34] The estuary is full of wildlife, not just fish. Sea otters and harbor seals call it home, as do herons, egrets, brown pelicans, and least terns, to name a few of the nearly 350 species of resident and migratory birds.[35] The estuary is a nursery for several fish, including a commercially valuable one, the English sole (*Pleuronectes vetulus*). Juveniles spend one to two years in estuaries like the Elkhorn Slough where they develop into adults before migrating back to deep Pacific waters. So, this fish would have to deal with estuarine hypoxia regardless of the season or month. Ecologists looked at fish abundance and diversity from 1988 to 2012 when dissolved oxygen varied in the Slough due to nutrients and to El Niño and its antipodal, La Niña.[36] Both are phases of the El Niño Southern Oscillation, which affects oxygen levels by modifying ocean temperatures and the upwelling of oxygen-poor waters.

The study found that oxygen has a big impact on fish in the estuary that translates into a big impact in offshore waters plied by commercial fishermen. Abundance of English sole was three times higher in the Slough, and fish diversity nearly 50 percent higher, when dissolved oxygen was at normal levels than in years when oxygen was below normal. Fish diversity as well as the abundance of English sole and of another flat fish, speckled sanddab (*Citharichthys stigmaeus*), were lower in the upper section of the estuary where dissolved oxygen was most depleted. The authors of the study calculated that for the entire estuary, hypoxia could lead to a loss of about 18,000 English sole and 7000 speckled sanddabs in a year. The death of English sole juveniles in the estuary should mean fewer adults to be caught by fishermen offshore. That was confirmed by trawl surveys conducted by the National Marine Fisheries Service, although the effect was delayed by a year or two because that is the amount of time the fish spends in the estuary as juveniles. The Elkhorn Slough can have such a big impact on an offshore fishery because it is the only estuary along 350 kilometers of California coast. Low dissolved oxygen in offshore waters where adult English sole live can also have a negative effect. But the biggest impact occurs in the estuary. Estuarine dead zones mean fewer flatfish and other fish that use these waters as nurseries.

I'll end with one final example of hypoxia shaping the distribution of commercially important fish. Analogous to the Fertile Fisheries Crescent abutting the Gulf dead zone, the eastern tropical Atlantic and Pacific Oceans have low-oxygen deep waters, while closer to the surface, in well-oxygenated waters, swim skipjack tuna, yellowfin tuna, and billfish.[37] The long-reach of the hypoxic deep waters is illustrated by looking at the depth billfish live in the Atlantic Ocean versus in the Pacific. In both oceans, the fish mostly live in the upper 50 meters, but the time they are detected in deeper waters differs in the two oceans. In the eastern tropical Atlantic Ocean, billfish spend 25 to 35 percent of their time below 50 meters, while Pacific billfish rarely go below 50 meters (1 to 5 percent). Fish ecologists believe that Pacific billfish are blocked from swimming long in deep waters because dissolved oxygen is lower and hypoxia more extensive in the eastern tropical Pacific Ocean. What's more, the habitat of these fish is slowly being squeezed by the expansion of hypoxic deep waters, what marine ecologists call "habitat compression." For reasons I'll explain in Chapter 9, the lowest depth with enough oxygen for these fish has been shallowing at about one meter per year in the northeastern tropical Atlantic Ocean; since the 1960s, billfish have lost about 15 percent of their habitat where the water has the right temperature and enough oxygen. Low-oxygen water is also creeping up closer to the surface in the eastern tropical Pacific Ocean. From tagged fish followed by satellites, habitat compression has been documented for blue marlin and sailfish. With less space to roam, billfish are more easily caught by sportsmen and accidently by fisheries targeting swordfish. The main threat to these fish is overfishing, but it's made worse by hypoxia.

So, the effect of hypoxia on fish is complicated. Some fish and fisheries may be affected, others may not. Even if the growth of a fish is slowed by low-oxygen waters, the effect may not show up in landings because those depend not only on the fish but also on the response of the fishermen and the rest of the fishing industry. Regardless of an effect on landings, fishery management has to take hypoxia into account in order to correctly interpret this year's data before properly setting fishery quotas for the next year and beyond.[38] A regulatory agency would set next year's quota differently if it knew a large harvest was due to a favorable environment for fish rather than higher catchability caused by hypoxia. It is true hypoxia doesn't seem to affect fish like mackerel and tuna that live near the surface far from the bottom where hypoxic waters lie. Tommy Pellegrin and other charter boat captains are likely to continue to haul in fish from the Gulf of Mexico regardless of the size of the Gulf dead zone. However, landings of benthic fish like flounder tend to decrease as hypoxic waters spread. In places like Elkhorn Slough, we see the negative impact of hypoxia on fish diversity and the landings of fish that use estuaries as nurseries. Shrimpers shouldn't be as sanguine as Captain Pellegrin, given the results of the Marty Smith study. The study illustrates the myriad of factors obscuring links between hypoxia and fisheries. It also shows the complications of proving hypoxia has an impact on fish harvests. You may get tired of hearing scientists say "it's complicated." I do too. But we cannot ignore dead zones just because the effect of hypoxia is complex.

Uncharismatic Animals

Perhaps because I study marine microbes and don't like seafood, I find the focus exclusively on commercially valuable fish and shrimp often too narrow and shortsighted. I see why showing the impact of a dead zone on fishery landings and prices is a powerful argument to do something about the problem, but I don't buy the converse: that if landings and prices are unperturbed, a dead zone isn't a problem. There is more to the ocean than what we can put on our dinner plate. It's the full diversity of marine life that evokes our awe and adoration, not just a few prized fish. Even if you only care about what you can catch, eat, or admire, you should be concerned about how hypoxia affects the many other less charismatic species that make up the foundation of marine food chains. These small uncharismatic animals low in the food chain feed the charismatic animals favored by anglers and lovers of seafood and nature. It's like my neighbor complaining about the peeling paint on my house while I was much more concerned about the house's century-old foundation, which consisted of only a stack of bricks in each corner.

We've already discussed one example of hypoxia affecting uncharismatic but important animals. The first victims of hypoxia are grubby polychaetes and other benthic invertebrates, which serve as food for flounder and other fish hugging the bottom and for shorebirds flying overhead. When bottom waters turn hypoxia, many fish and all birds can escape, but the benthic invertebrates are mostly wiped out, leaving a denuded habitat behind.

Hypoxia also affects the food chain in the water column. To explain this effect, I need to review the main parts of a typical aquatic food chain (Fig. 8.2). It starts off with growth or primary production by algae, which in turn are eaten mostly by small herbivores, the zooplankton. You may have eaten one type of zooplankton, krill, if you've had okiami in a Japanese restaurant or taken krill oil pills to get your daily dose of omega-3 fatty acids. Krill is food for baleen whales and penguins in Antarctica and fish in many oceans. Other zooplankton are equally nutritious for adult and juvenile fish. The aquatic food chain doesn't work as it should when zooplankton are messed up by hypoxia. There are fewer of them over the entire water column when the bottom layer is hypoxic, suggesting lower growth, greater mortality, or more emigration to well-oxygenated waters.[39] Hypoxia also screws up vertical

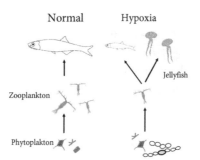

Normal   Hypoxia

Zooplankton

Jellyfish

Phytoplakton

Figure 8.2 Typical food chain without (normal) and with hypoxia. Phytoplankton include algae and cyanobacteria. With hypoxia, the zooplankton community shifts to smaller species, which selects for smaller fish predators. Because of differences in their tolerance of hypoxia, fish decrease and jellyfish can become more common as dissolved oxygen disappears.

migration. Many zooplankton migrate up the water column at night to feed and down the water column during the day to avoid visual predators—those that need to see their prey before striking. Migration becomes complicated if not impossible when bottom waters have too little oxygen. If restricted to the surface layer because of hypoxia, another example of habitat compression, zooplankton are more easily eaten because they can be seen more easily by fish and other visual predators, and they are more concentrated in a smaller space.

What's more, low oxygen levels may change the size and types of zooplankton in a habitat. This will take a little explaining. The first step in the argument is to note that zooplankton depend on diffusion to take up dissolved oxygen from the surrounding water. A large zooplankton has more surface area than a small one to take up oxygen by diffusion, but it also has a bigger body that needs to be supplied with the gas. As a net result, being big is a disadvantage with regard to gas exchange; a large zooplankton has more difficulty than a small one getting enough oxygen when concentrations are low. To see why, consider a spherical zooplankton. There are none, but the argument is the same regardless of the zooplankton's exact shape. The surface area of this spherical zooplankton depends on the square of its radius while the body size or volume depends on the cube. So, as size increases, the ratio of oxygen intake (area) to oxygen use (body size) decreases, meaning it's best to be small when oxygen is low. The theory is supported by data for zooplankton in Tokyo Bay, Chesapeake Bay, and the Gulf of Mexico. In these waters, as hypoxia increased, so too did the numbers of small zooplankton.

What's the big deal? In the plankton's world, predation depends on size. Smaller zooplankton are eaten by smaller carnivores, and those animals in turn are eaten by smaller predators. The effect ripples up the food chain. Even if a fish or another carnivore could eat the smaller zooplankton, it would have to eat more of them to get the same amount of food. It's an example of how hypoxia rearranges food webs.

More than just a rearrangement, the replacement of fish by jellyfish would overhaul food webs. As dead zones expand, jellyfish may become more abundant at the expense of fish. Denise Breitburg and her team working in the Chesapeake Bay found that ctenophores, better known as comb jellies, are more tolerant than fish of low oxygen levels.[40] Denise mentioned to me that she and her colleague MaryBeth Decker ended an experiment after their patience ran out and before a ctenophore succumbed to very low dissolved oxygen.[41] As oxygen levels decrease, jellyfish can continue to be active and eat zooplankton, the prey favored by some fish. So, even if fish escape the direct effects of hypoxia, they would have less food and would be pushed out by jellyfish. In spite of their name, jellyfish are far from being fish. Jellyfish are in the phylum Cnidaria along with sea anemones and corals, whereas fish are in the Chordata, the same as *Homo sapiens*. The contribution of jellyfish and fish to the food chain also greatly differs. Being mostly water, jellyfish are less nutritious than fish and other marine animals; they have been called the dead end of food chains. Then there are more direct effects of jellyfish on us, including the stinging of swimmers and the clogging of fishing nets and cooling systems of power plants. Even though jellyfish have an ethereal beauty, their takeover of the oceans strikes me as a particularly creepy result of hypoxia.

If hypoxia kills or screws up the base of the food chain, those uncharismatic benthic worms and zooplankton, you'd think that fish and other animals higher up on the food chain would also be affected. But remember the relationships between fish, fisheries, and hypoxia are complicated. Fishery landings may or may not decline with dead-zone size. Regardless, we should be concerned about hypoxia causing cracks in the base of the food chain. My house probably would have stood another century with only bricks holding it up, but I feel better knowing it now has a proper foundation (and my neighbor is happier that the house has been painted). The Covid-19 pandemic provides another analogy. Doctors have used the term "silent hypoxia" to describe coronavirus-infected patients who are having no trouble breathing yet have dangerously low oxygen levels.[42] Even if the patient—or habitat—seems fine, hypoxia never should be ignored.

Dead benthic polychaetes, shrinking zooplankton, and jellification of the oceans are dead-zone casualties that go uncounted except by the specialist. That's the basic challenge in convincing the public that hypoxia is a serious problem. Except for dead fish that float to the surface or wash onto beaches, you can't see the havoc unless you happen to go diving in waters turned hypoxic, as Denise did when winds pushed deep dead-zone waters onto the west-side shallows of the Chesapeake Bay. There she saw dead anchovies and blue crabs climbing on top of each other, trying to escape the hypoxic waters. But the rest of us have little chance of witnessing dead-zone impacts firsthand. With the help of photos and video clips, it's easier to grasp the remains of environmental disasters on land, such as the blackened branches and tree trunks left behind by forest fires in California and Australia. It is impossible for an underwater photograph to capture the scale of a dead zone's destruction because hypoxic waters are so murky and there is so little to see. Imagine the sadness if 15,000 square kilometers—the average area of the Gulf dead zone—of Yellowstone National Park burnt down, or the uproar if 15,000 square kilometers of farmland in Iowa were bulldozed under. Somehow dead-zone scientists have to find a way to evoke similar sadness and uproar over the world's dead zones.

9

Dead Zones in the Oceans

As if there weren't enough things to worry about in the late 1960s—between assassinations, wars cold and hot, earthquakes, hurricanes, social unrest, and political upheaval—some ecologists worried about oxygen.[1] In 1970, the great oceanographer Wally Broecker pointed out that "all grocery lists of man's environmental problems" have an item about the supply of oxygen.[2] He wasn't referring to the supply in water for aquatic organisms but rather the supply in the atmosphere for us and other air-breathing life on land. Some ecologists in the 1960s wondered if DDT or other chemicals could inhibit photosynthesis and thus suppress oxygen production and lower atmospheric oxygen levels. Broecker concentrated on another human activity debated at the time: the burning of fossil fuels. Coal, oil, and natural gas started as plant and algal organic carbon that was deposited during the Carboniferous period about 300 million years ago. As that organic carbon was made by oxygen-evolving photosynthesis and then was buried before it could be degraded, oxygen accumulated in the atmosphere to high levels, perhaps even higher than today's 21 percent. In the late 1960s, it was feared that burning of this fossil organic carbon would return us to a past geological time with significantly lower oxygen levels. However, Broecker correctly calculated that using up even all fossil carbon reserves known at the time would not deplete the atmosphere of oxygen. He said, "There are hundreds of other ways that we will hazard the future of our descendants before we make a small dent in our oxygen supply."

Broecker was mostly right but not about his implication that we don't have to worry about any dents in our oxygen supply. The dent we are making, albeit very tiny, has worrisome implications for the oceans and the rest of the biosphere.

The loss in atmospheric oxygen is so small that it takes a method sensitive enough to detect a change as tiny as five oxygen molecules among a million other gas molecules.[3] Studies using this method have found that the atmosphere is losing about 19 oxygen molecules out of every million air molecules each year.[4] Although the amount seems minuscule, and we're not in any danger of ever running out of the gas, our atmosphere is definitely losing oxygen. This small loss has some big things to say about the carbon cycle. Ralph Keeling, the scientist who first came up with the method, was able to match up the loss with the increase in atmospheric carbon dioxide first measured by Keeling's father, Charles Keeling, in 1958 on the Big Island of Hawai'i.[5] The yearly loss of those 19 oxygen molecules has been used to explore global rates of primary production and the fate of carbon released from the burning of fossil fuels. The loss of atmospheric oxygen is another sign that the atmosphere is gaining carbon dioxide and that our climate is changing as a result.

The yearly loss of the 19 oxygen molecules from the atmosphere is also connected to the loss of many, many more oxygen molecules from the open oceans and some coastal waters. While the decline in atmospheric oxygen is a symptom of a much bigger problem, climate change, the loss of dissolved oxygen is arguably the more immediate problem for life in many oceanic habitats.

The term used by oceanographers for this loss of oxygen is "deoxygenation." It's a word only a scientist would appreciate, I suspect. I have to think about how to pronounce it, to make sure I get all of the syllables in. Oceanographers working on deoxygenation probably like the term dead zone even less than do ecologists working on waterbodies like lakes or the Baltic Sea. Despite being ungainly, the word deoxygenation can be called into service to describe even small losses of dissolved oxygen well short of dead-zone levels. Make no mistake, more of the ocean is turning hypoxic and anoxic. But even vaster regions of the oceans are losing some oxygen.

Deoxygenation of the Open Oceans

It is challenging to detect changes in oxygen over time in the open oceans far from land. The problem is separating the signal—a change over years—from the noise: the daily ups and downs and the variation due to the mixing of different water masses with different dissolved oxygen content. The solution is to measure oxygen over a long time—the time-series approach that was crucial in revealing the cause of dead zones in the Gulf of Mexico, the Adriatic Sea, and the Baltic Sea. In addition to the obstacles faced by time-series programs in coastal waters, programs in the open oceans have additional costs and logistical hurdles in studying a habitat far from home. One of the few regions in the open oceans with lots of oxygen data over a long time is the subarctic North Pacific Ocean. Many oceanographic studies have focused on one spot, Station Papa, often called just Station P, at 50 degrees north, 145 degrees west, about 700 miles south of the port town of Seward, Alaska.

First called Station Peter, Station Papa was established by the US Navy in 1943 to collect weather data during World War II.[6] Although the Navy also occasionally measured the ocean's temperature, most of the early ocea-nographic work was done by Canadians who took over the station after the Americans left in June of 1951. Canada sent out weather ships built especially for meteorological work, but the Canadians also did some oceanography on the side. When the weather ships stopped going to the station in August 1981, fortunately the oceanographic time-series work was continued by scientists at the Institute of Ocean Sciences based in Sidney, British Columbia.

Station Papa is famous among oceanographers and infamous among those who have sailed there. It's famous because it's the site of many important studies in oceanography, including one of the first showing that biological production in some oceans is limited by iron, not nitrogen or phosphorus nutrients. The station is infamous because of its huge swells and deep troughs, its big white-capped waves, and unceasing winds. It is overcast and dark even on a storm-free day. I was reminded of my first cruise to Station Papa in June 1987 when I stumbled on a letter from Charles Darwin who was writing to his cousin about his experience on the HMS *Beagle*: "I hate every wave of the ocean with a fervor, which you, who have only seen the green wa-ters of the shore, can never understand."[7] Often seasick while on the *Beagle*, Darwin should have taken scopolamine. Pills and patches of the drug helped me survive my four month-long trips to the station.

Early work by the Canadians on the weather ships, probably without any seasickness meds, provided the first data points that eventually revealed the loss of dissolved oxygen at Station Papa. The first data points in July 1956 came from water samples collected at several depths and analyzed using the Winkler method on the ship. A similar approach is still used today, but many, many more data points are now taken from oxygen meters tethered to the ship or mounted on Argos floats and gliders; the latter are torpedo-shaped, autonomous vehicles that measure several oceanic properties as they weave slowly up and down through the water column. After 50 years, oceanographers had enough data to see a startling pattern emerge: Station Papa was losing oxygen.[8] The loss is measurable at all depths, but it's espe-cially large in a layer of water between 150 and 200 meters. Between 100 and 400 meters and the years 1956 and 2006, oxygen declined by 22 percent.

Deoxygenation was also measurable in nearby Okhotsk Sea and close to the North America coast. Arguably more disturbing than the deoxygenation of the upper waters at Station Papa is the expansion of its hypoxia zone. Yes, below the surface layer in the middle of the North Pacific Ocean is a dead zone. It was very deep, about 400 meters in 1956, but by 2006, it had shoaled to 300 meters. No published study has looked at what's happened since 2006, although I did find some data collected in 2014 by the Canadians and other data collected a year later at a site about 670 kilometers to the northwest of Station Papa.[9] What these and other data indicate is that the subarctic Pacific is continuing to lose oxygen at a rate of about 3 percent per decade.[10]

Rather than dead zones, oceanographers call these layers of hypoxic waters "oxygen minimum zones," or OMZs. To earn the label, oxygen concentrations have to be even lower than the level seen in a dead zone. In oceanography, hypoxia is defined as water with less than 63 micromoles of oxygen per kilogram of water, which is about the same as the definition, 2 milligrams per liter, I gave in the Prologue. (I won't try to explain why oceanographers use micromoles. They use a measure of weight, kilograms, rather than a measure of volume, liters, to avoid the complications caused by the compression of water by pressure.) An OMZ is usually defined as having less than 20 micromoles of oxygen per kilogram. Although the precise level isn't important, it is worthwhile remembering that an OMZ has less oxygen than many of the dead zones we've talked about so far.

In addition to the Station Papa OMZ, there are several other OMZs around the world (Fig. 9.1). The biggest ones include two in the eastern tropical Pacific Ocean: one north of the equator, streaming west from Mexico and Central America, and the other south of the equator, mostly off the coast of Peru. Another big one is in the Arabian Sea and the Bay of Bengal, two northern lobes of the Indian Ocean. Paralleling the OMZs in the eastern tropical Pacific Ocean are the low-oxygen waters in the tropical Atlantic Ocean, extending off the coast of Africa. The tropical Atlantic is not an OMZ by the standard definition of 20 micromoles of oxygen per kilogram, but water at some depths are well below the cutoff for hypoxia. Off the coast of Namibia in southern Africa, oxygen is so low that hydrogen sulfide can be produced, forcing a "lobster walkout" as mentioned before. Oxygen is low in these regions because algal growth and production of organic material is high, stimulated by nutrient-rich deep water driven to the surface by wind—an upwelling. As with other dead zones, the algal material is either used by zooplankton and the rest of the marine food chain, or it sinks to deep waters

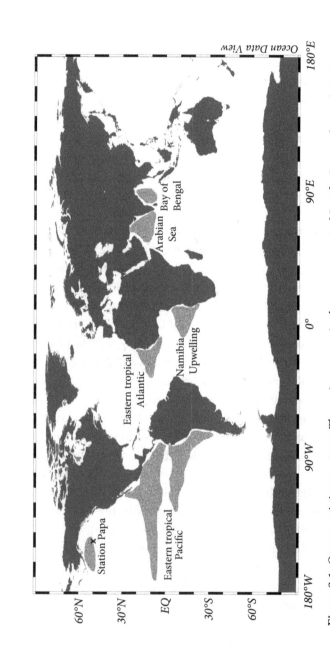

Figure 9.1 Oxygen minimum zones. The main ones are in the eastern tropical Pacific Ocean, the Arabian Sea and the Bay of Bengal, and off the coast of Namibia in southern Africa. The eastern tropical Atlantic Ocean has hypoxic waters, but its oxygen level is higher than in OMZs. Oceanic deoxygenation was first documented at Station Papa.

where it is degraded, using up dissolved oxygen in the process. Also like some coastal dead zones such as the Gulf of Mexico, OMZs are associated with rich fisheries, the most notable being the anchovy fishery off of Peru's coast.

The OMZ at Station Papa is not the only one expanding. On a global basis, from 1950 to today, the volume of OMZ waters in tropical waters has increased by about 20 percent.[11] But each OMZ has its own story. The thickness of the tropical Atlantic low-oxygen layer, for example, has expanded by 85 percent,[12] while the volume of OMZ water in the eastern tropical Pacific has increased by about 7 percent.[13] The latter percentage may seem small, but it's applicable to a large area and represents a huge increase in poorly oxygenated waters, equivalent to the area of Australia. Along with the expansion in area and volume, OMZs have become more intense; oxygen levels in the OMZ and in surrounding waters have dropped, by as much as 50 percent in the eastern tropical Pacific.[14] Deoxygenation can be more or less severe, depending on natural ups and downs over time of oxygen and everything else in the oceans. A big source of these swings is the Pacific Decadal Oscillation, an El Niño-like variation in the atmosphere and climate with impacts on the oceans and beyond. During the positive phase of this Oscillation, dissolved oxygen levels drop even more, and OMZs expand at a brisker rate.[15]

But that's a detail, perhaps of interest only to oceanographers. Let's not let the Pacific Decadal Oscillation obfuscate the bigger question: why are the oceans losing oxygen?

Global Warming is Ocean Warming

Scientists had observed low oxygen when they started to measure gases in the oceans in the 1800s. Early oceanographers first saw low levels in 1874 during the HMS *Challenger* expedition,[16] the first project to systematically explore many of the world's oceans. The *Challenger* set sail when the basic chemistry of gas exchange was still being worked out and before Ludwig Winkler came up with the Winkler method for measuring dissolved oxygen. Although the author of this part of the *Challenger* report, Professor William Dittmar, had the ingredients for the right answer, he said he wasn't successful in connecting the low values to "natural causes" and thought some of the low values were due to incomplete extraction of gases from seawater, a necessary step in measuring oxygen at the time. A few decades later, in January 1922, now equipped with the Winkler method, a Danish oceanographer, Johs

Schmidt, discovered waters off of Panama with very low oxygen.[17] The work was done as part of the Danish *Dana* expedition supported by the Carlsberg Foundation, which got (and still gets) its money from the Carlsberg Brewery. Schmidt was so surprised to see the oxygen data that he used italics, rarely seen in scientific papers except for species names, to make sure the reader got his message: "the water here *contains practically no oxygen at all.*" He mentioned the *Challenger* finding low oxygen in the Caribbean Sea and other reports of low levels in the Baltic and Black Seas. But at the time, his estimates for dissolved oxygen in an open ocean were the lowest. He had discovered the eastern tropical Pacific OMZ.

While low oxygen levels in the open oceans aren't new, what is new, probably since the mid-20th century, is the expansion of the OMZs and the loss of oxygen in other oceanic waters. Once again, we humans look like the suspects. One report suggested a type of nutrient pollution, iron and nitrogen chemicals carried by the winds, has stimulated algal production and oxygen depletion in the tropical Pacific.[18] Most oceanographers think, however, that a large part of oxygen loss in the open oceans is explained by temperature.

The oceans have taken on heat, gotten warmer, and as a result have lost oxygen. The average oceanic temperature has increased by about 0.1 degree Celsius per decade since 1951, and the total heat content in the global ocean has increased by about 10 times over the last 50 years.[19] The oceans have been warmed by the heat trapped by atmospheric carbon dioxide and other greenhouse gases. In the absence of the oceans soaking up so much heat (over 90 percent trapped by the atmosphere), the climate we experience on land would be warming even more than at its current pace. Global warming means ocean warming.

Although it has slowed down the overheating of terrestrial habitats, ocean warming comes with costs. A basic law of chemistry dictates that the solubility of oxygen and other gases in water decreases with increasing temperatures; warmer waters can hold less oxygen. The solubility law is a good start at explaining why the oceans are losing oxygen. It explains deoxygenation of the upper 100 meters of the oceans,[20] but it accounts for only about half of the oxygen loss in deeper waters down to 1000 meters.[21] Overall, the solubility effect of temperature explains only about 20 percent of deoxygenation of the open oceans.

Temperature is connected to three other processes that explain oceanic deoxygenation. One is biological. While production of oxygen by photosynthesis is generally thought not to depend on temperature, the consumption of oxygen by microbes is well known to vary with temperature. The cold temperature of a refrigerator slows down microbes enough to keep a turkey sandwich edible for a few days, whereas microbes would quickly ruin that sandwich left outside at room temperature. This temperature effect is encapsulated in "Q_{10}," which is the relative stimulation of metabolic activity by a warming of 10 degrees Celsius. Most experiments have found that raising the temperature by 10 degrees doubles the rate of oxygen consumption, meaning the Q_{10} is usually equal to about two. However, a study in the Sargasso Sea estimated a much greater Q_{10} of 3.6,[22] and a slightly higher Q_{10}, 3.7, was found in the Indian Ocean.[23] If those Q_{10} values are correct, then oxygen consumption would increase by nearly 30 percent if the oceans warm by 2 degrees Celsius, which is already occurring in some locations. Although we know a lot about Q_{10}, we don't know how much of today's oceanic deoxygenation can be ascribed to higher oxygen consumption in a warming ocean. While it is fairly easy to see how microbial activity varies with temperature in a controlled experiment, it is much more difficult to predict how the activity-temperature relationship plays out in a real ocean. This question has been explored in much more detail on land where the answer determines how much carbon dioxide may be released by global warming of soils.

The second temperature-related process contributing to oceanic deoxygenation involves stratification and the mixing of water masses. We saw before that stratification in dead zones like the Gulf of Mexico and the Baltic Sea is set mainly by salinity. Less dense, low-salinity water sits on top of denser, salty water. Temperature is more important in setting stratification in the open oceans, especially with global warming. Stratification in the oceans is from warmer, lower-salinity surface waters sitting on top of colder, more salty deep waters. With climate change, stratification in the open oceans could become stronger because the surface layer is warming faster and becoming even lighter than deeper waters. Because it lessens exchange between surface and subsurface waters, stronger stratification would have two competing effects on oxygen levels in deep layers. It could lead to lower oxygen in deep waters. The diminished exchange means less well-oxygenated surface water is mixed down into deeper layers, giving microbes more time to deplete oxygen. An oceanographer would say that stronger stratification increases the "ventilation age" of deep waters—it lengthens the time since these waters were last exposed to the atmosphere and high oxygen levels. Another effect of stronger stratification could lead to the opposite result—higher oxygen concentrations because stratification impedes the supply of nutrients into the surface from deep waters. Without those nutrients, algal production decreases as does the supply of oxygen-consuming organic material to deep waters, allowing oxygen levels to remain high. Oceanographers don't know which of the two competing effects will win in the future. The winner likely varies with the oceanic region and time.

Temperature-drived changes in ocean circulation is the third process that could lower oxygen content.[24] Physical oceanographers have hypothesized that the eastern tropical Pacific Ocean is losing dissolved oxygen because it is receiving less oxygen-rich water from elsewhere when winds that drive the transport of oxygen-rich water diminish. Transport may have declined already by as much as 30 percent according to one estimate. A similar mechanism explains about half to two thirds of the oxygen lost from the St. Lawrence Estuary in Quebec. In this case, oxygen levels in the estuary have declined with lower flow of oxygen-rich water from the Labrador Current. Some of these changes in circulation may be part of natural swings in our climate like the Pacific Decadal Oscillation and El Niño. But other changes are thought to be due to global warming.

I implied that oceanographers were startled to see the loss of oxygen at Station Papa, but that's a bit of an exaggeration. They were forewarned. Knowing the planet was warming and armed with information about the temperature-dependent solubility of gases, climate-change modelers had predicted the oceans would undergo deoxygenation. That wasn't the first time models helped us to understand low-oxygen problems. One of the first models of a dead zone was quite unlike the computer models of today. In 1976, the US Army Corps of Engineers built a scale model made of concrete and filled with real water, to depict the Chesapeake Bay.[25] The model took up a 9-acre (3.6 hectares) building and cost $14 million. It was a flop. It failed to get the vertical scale right, which was needed to adequately replicate the Bay's hydrography. As computing power soared in the late 1970s, the concrete model became obsolete, and it was abandoned in 1983.

The Chesapeake scale model was nothing compared to the one the Corps built to simulate the Mississippi River basin.[26] This scale model covered 200 acres (81 hectares) near Clinton, Mississippi, although the heart of the model—representing the basin from Hannibal, Missouri to Baton Rouge, Louisiana—took up only 20 acres. The first parts of the model were built by German and Italian World War II prisoners-of-war starting in 1943, and the model's last components were finished in 1966. But enough of the model was available for experiments to be run in 1949. The model was put to the test dramatically in 1952 when flooding threatened two cities, Omaha and Council Bluffs, along the Missouri River, which joins the Mississippi near St. Louis, Missouri. The model identified the levees that needed shoring up, and those recommendations were applied immediately by the two cities, saving an estimated $65 million in damages. The model continued to be used until 1971, when it too was killed off by computers. Remnants of the model are now part of Buddy Butts Park.

Still another model of the Mississippi River was built by a mechanical engineer turned entrepreneur.[27] In 1928, Carroll Livingston Riker installed a scale model—complete with dams, valleys, and levees that held back running water—in a basement room of the US Capitol in Washington, DC, to advertise his plan to control flooding along the Mississippi. More risible was Riker's scheme to build a 320-kilometer jetty extending out from Newfoundland, which he calculated would redirect the Gulf Stream north of its current course and clear ice-clogged harbors in Canada and shipping lanes of icebergs in the North Atlantic. Neither plan went anywhere.

Even the 200-acre model of the Mississippi River basin is no match for today's computer models. Relatively simple statistical models can accurately predict the size of the dead zone in the Gulf of Mexico and Lake Erie, and more sophisticated models faithfully reproduce the patchiness of hypoxic waters in dead zones. Computer models have also gotten right the amount of dissolved oxygen we know has already been lost from the world's oceans, about 2 percent overall since the 1950s.[28] They predict that as much as 7 percent will be lost by 2100. If those percentages seem small, remember even small changes in dissolved oxygen can affect marine biota, and the change is much greater in some oceans like the eastern tropical Pacific with its large OMZ.

Unfortunately, it is in the tropical OMZs that the models haven't done a good job so far. In contrast to the known decrease, the models predict that dissolved oxygen in OMZs should increase. There are several possible reasons for why the models are wrong. They may be wrong about equatorial jets and other currents that mix oxygen-replete and oxygen-depleted waters. Most of these models have simplistic depictions of how microbes produce or use oxygen, but it is not clear which depiction, if any, is to blame for the models' failure. Where models fall short may be informative about what controls oxygen levels in the oceans. As the statistician George Box said, "all models are wrong, but some are useful."[29]

Deoxygenation of Coastal Oceans

Oregon fishermen first noticed something was wrong in 2002 when they pulled up their crab traps and found only dead or dying animals.[30] Dead fish and invertebrates washed ashore that year, and videos from remotely operated submersibles found more death where fish and crabs usually abound. From another submersible in August 2006 came images of white mats of bacteria, some of which are known to consume hydrogen sulfide, evidence of the complete absence of oxygen—anoxia, not just hypoxia. Based on records going back five decades,[31] low to no-oxygen waters had not been observed in these waters before 2000. Further north, near Point Grenville of Washington state's Olympic Peninsula, dead fish washed onto a beach popular with members of the Quinault Indian Nation reservation.[32] Tribal leaders and elders don't know of any stories about such a notable event before the first fish kill occurred in July 2006. Now the low-oxygen waters have become a yearly summertime feature of the Oregon and Washington coast.

The odd thing about this coastal dead zone is that a main ingredient, a river, is missing. Other dead zones in coastal waters and seas are fed by rivers. The Black Sea has the Danube and the Dniester and six other rivers, and the northern Gulf of Mexico has the Mississippi and Atchafalaya. The Neva, Vistula, and Oder are just three of hundreds of rivers that flow into the Baltic Sea. It is true that the border between Oregon and Washington is the Columbia River, but its plume heads offshore when it enters the Pacific Ocean and is seaward of the coastal hypoxia. It can't be blamed for the dead crabs and fish. Oregon and Washington have other rivers, but they are small and don't flow past major metropolises or vast croplands. These rivers, even the Columbia River, with the city of Portland on its banks, are puny when matched up against the Pacific Ocean.

The Oregon and Washington coasts are not the only ones losing oxygen. Levels of the gas are going down along the entire west coast of North America; already dissolved oxygen at 250 to 400 meters has declined by 40 percent since 1998.[33] These waters are part of the California Current, which starts off of British Columbia; flows south along the coast of Washington, Oregon, and California; and eventually runs past Mexico's Baja peninsula. It's an example of what oceanographers call an "eastern boundary current;" although on the west coast of North America, the California Current is on the east side of the Pacific Ocean. The South American version of the California Current is the Humboldt Current, while Africa has the Benguela Current. These eastern boundary currents have always attracted a lot of attention. Although covering only 1 percent of the ocean, they account for about 20 percent of the global wild fish harvest.[34] The fish are there because algal production is high, fueled by the upwelling of nutrient-rich, deep water to the surface. I mentioned the anchovy fishery of Peru already. Today some of these eastern boundary current systems are also being studied because they are losing oxygen.

Deoxygenation of the California Current was initially hard to explain, but now we have some good ideas about why these waters are losing oxygen. One reason is the water feeding the California Current, its "source water" to use the oceanographic term, is losing oxygen. The source water includes the subarctic Pacific, home to Station Papa, and the hypoxic waters of the OMZ to the south, in the eastern tropical Pacific. As these OMZs expand and the waters above and around them lose oxygen, the source water feeding the California Current also has less oxygen. Upwelling along the North American West Coast exacerbates the problem. The organic material from upwelling-stimulated algal production eventually draws down oxygen, already at low levels, even further. What's more, the winds driving upwelling are getting stronger, making the upwelling more intense. We're still not done. Along with stronger winds and an expanding OMZ are the other factors we discussed before that are causing oceanic deoxygenation: warming waters can hold less oxygen, and what is there is consumed at a higher rate by bacteria.

Nearly all of the suspects are now in place to explain the death of crabs and other animals along the northwest coast of North America. The animals are there in high abundance because of the upwelling, but the upwelled waters are becoming lethal as oxygen concentrations drop. Exacerbating the hypoxia effects, the upwelled waters are also becoming acidified (lower pH), another result of organic material degradation and higher carbon dioxide concentrations. Acidity can interfere with respiration and makes some animals more sensitive to low oxygen levels.[35] We need one more piece of the puzzle to explain the increasing fatality of marine animals along the Oregon coast: changes in the winds and currents. Stronger winds not only intensify upwelling, they also drive stronger currents that push hypoxic, acidic water closer to shore where it can do the most damage to marine life. The end result is the dead crabs in fishermen's traps and dead fish washing ashore on Oregon and Washington beaches.

Deoxygenation Winners and Losers

Just as some people profit from natural disasters, pandemics, and wars, some marine organisms may be winners as the oceans lose oxygen. Here it may be a good place for a reminder that dead zones aren't really dead. In addition to teeming with microbes, hypoxic waters can provide a refuge for animals like krill and some fish trying to escape visual predators during the day.[36] There are more winners. Invertebrates, such as vampire squid and mysid shrimp, now living in the low-oxygen waters of the California Current, are thought to benefit from the expansion of hypoxic waters, and likewise for the Humboldt squid, also called the jumbo squid (*Dosidicus gigas*; Fig. 9.2). As implied by the name, this squid can be huge, up to 2.5 meters and 50 kilograms.[37] Depending on the animal's life stage, it hunts in packs of over a thousand animals, and is capable of swimming 5 to 25 kilometers per hour. The northern limit of this squid used to be northern Baja California, but now it is found as far north as southeastern Alaska, perhaps because low-oxygen waters have expanded along the North American west coast.[38] The squid may be doing well because it tolerates low oxygen levels and is adept at finding prey in the layer above hypoxic waters. The shoaling of hypoxic water is squeezing the habitat for the squid's prey, concentrating them and making them easier targets for the squid. It's another example of habitat compression.

Figure 9.2 The Humboldt squid (*Dosidicus gigas*). Drawing by G. Pfeffer (1912), provided by Richard Young.

As the Humboldt squid moves into new territory with the spreading of low-oxygen waters, it undoubtedly does not come alone. The range of other animals is also likely increasing. We know a fair amount about the squid's physiological response to hypoxia and about its ecology, in part because it is a target of commercial and sport fisheries. It is more difficult to make conclusions about the biogeography of the many understudied marine animals. When one of these is found in a new location, marine biologists are reluctant to proclaim that its range has expanded. Perhaps we hadn't looked hard enough to detect it before. Even so, along with the Humboldt squid, the vampire squid, and mysid shrimp, other organisms probably profit from expanding low-oxygen waters.

For every Humboldt squid, however, there are likely ten or more losers that are selected against by deoxygenation. The most obvious losers show up dead in crab traps or on beaches. There are other less dramatic but more systematic and disturbing downward trends. Since 1951, as the California Current has lost oxygen over the years, pelagic fish have declined in those waters by over 60 percent.[39] Increasing hypoxia in the oceans is expected to contribute to a decline in the diversity of marine animals, and the expansion of OMZs is thought to be one reason why the abundance of marine predators has already decreased by 10 to 50 percent.[40] Even if a fish or invertebrate can survive hypoxia, it may be more susceptible to diseases and parasites.[41]

Marine life is threatened by other global-level threats that are co-occurring with and connected to oceanic deoxygenation. The oceans are not only losing oxygen and gaining heat, they are becoming more acidic as they suck up more and more carbon dioxide from the atmosphere. The combined effect of these "multiple stressors" is greater than the effect of one stressor acting alone, just as people with chronic lung diseases and other health problems are more vulnerable to the Covid-19 virus. I already mentioned how acidity makes some organisms less tolerant of low-oxygen concentrations. A warming ocean likely has an even bigger effect. An organism may need more oxygen in a warmer ocean if its metabolism is stimulated by the higher temperature, making it more sensitive to low levels of oxygen. Some ecologists have speculated that if oceanic waters warm by 4 degrees Celsius, benthic animals would be reduced by 36 percent.[42] One study concluded that oceanic habitats with suitable oxygen levels and a cool-enough temperature for normal organism activity could shrink, depending on the model and various assumptions, by 17 to 25 percent before the year 2100.[43]

The Nitrogen Cycle and Global Algal Production

Deoxygenation of the oceans would be worrisome enough if it only affected the animals that we can see swimming in the water or crawling along the bottom. But I think the scariest, most profound, and far-reaching impacts are on things we can't see: two reactions in the nitrogen cycle. One reaction produces the gas nitrous oxide (N_2O) and another nitrogen gas (N_2).

Nitrous oxide has a colorful history since it was first discovered in 1772 by an Englishman, Joseph Priestley, who also co-discovered oxygen. One of his compatriots and colleagues, Humphry Davy, did more of the early work on the nitrous oxide. In addition to trying out manure cocktails (Chapter 6) and discovering fish need dissolved oxygen, Sir Humphry tested nitrous oxide on a "healthy kitten of about five months old," a small dog, and himself.[44] He discovered its narcotic and mind-blowing effects, observing that while under the gas's influence, he felt that "Nothing exists but thoughts!—the universe is composed of impressions, ideas, pleasures, and pains!"[45] A popular and gregarious man, he shared the drug with his friends who included James Watt, Samuel Taylor Coleridge, Robert Southey, and other illuminati of the day. Gatherings of the English upper class imbibing the gas were called "laughing gas parties" because nitrous oxide can induce the giggles and other signs of euphoria.

There is a less humorous side to nitrous oxide. It's a potent greenhouse gas, 200 to 300 times more effective than carbon dioxide in trapping heat from sunlight. The level of nitrous oxide in the atmosphere has been increasing over the last century, mainly due to agriculture and the increasing use of nitrogen fertilizer, the largest source of the gas. Still, a lot of nitrous oxide comes from the ocean, about 25 percent of the global total,[46] and a large fraction of the ocean's contribution comes from low-oxygen waters in or near OMZs. Expansion of these waters by global warming adds more of this greenhouse gas and further exacerbates global warming. It's an example of "positive feedback." More nitrous oxide leads to more global warming, which leads to more oxygen loss, which leads to more nitrous oxide, and so on. In this case, the positive feedback is very much a negative.

Nitrous oxide is also involved in a potential "negative feedback" that may slow down loss of oxygen from the oceans, but this feedback has a downside. The release of nitrous oxide to the atmosphere means less fixed nitrogen useable by algae remains in the oceans. The other nitrogen cycle reaction affected by oceanic deoxygenation, the one that produces nitrogen gas, is another route, the main one, by which the oceans lose fixed nitrogen. This second reaction, the last step in denitrification, occurs only in the absence of dissolved oxygen. So, the expansion and the intensification of OMZ regions especially in the eastern tropical Pacific Ocean, where oxygen is already very low, will mean the oceans could lose more fixed nitrogen than in the past. With less fixed nitrogen, algal growth will decline in those waters limited by the nitrogen supply. The good thing is that less algal growth means less organic material that fuels oxygen consumption and loss of oxygen. It's a negative feedback. The downside is that lower algal growth, the engine that drives all aquatic food chains, could be a negative for all other organisms in the oceans. The amount of hard data to explore these issues is small compared to the size and complexity of the problem, so it is easier to speculate than to draw firm conclusions. But, it is clear that oceanic deoxygenation is disrupting the nitrogen cycle, which is essential for ocean biota and the rest of life on the planet.

Wally Broecker was right, we don't have to worry about suffocating and running out of oxygen. We won't run out of breath anytime soon. But we should worry about the loss of those 19 oxygen molecules from the atmosphere for other reasons: it is a sign of a huge problem, the burning of fossil fuels and the relentless buildup of greenhouse gases in the atmosphere. Loss of oxygen from the oceans could be taken as just another sign of this huge problem, but it is more than that. It is also stressing marine life and potentially the rest of the biosphere.

10

Reviving Dead Zones

You just saw in Chapter 9 that the open ocean and some coastal waters are losing oxygen because of global warming. We've already seen that global warming and other aspects of climate change are contributing to oxygen depletion of many water bodies. Climate change and dead zones share other connections and similarities. Climate change has been called a "super wicked problem"[1] because it is so complex and wickedly hard to solve, whereas the hypoxia problem has been called just "wicked."[2] Whether super or not, reviving dead zones is very difficult, with so many moving parts and complicated actors, that it may be hard to know where to start. I'd like to begin with something very simple: a small pond on the twelfth hole at Kings Creek Country Club where my wife and I play golf.

The pond doesn't come into play unless you hook or pull the ball badly or do some combination of both. Like many bad golfers, I tend to slice everything to the right, so I don't think about the pond. Hole 13 heads back on the other side of the pond, so again it is on our left but now more in play. Although I worry more about hitting my drive out of bounds on the right, my wife, who is a much better golfer than me, has to watch out for pulling her drives to the left where lime-green water waits. The pond receives nutrients from heavily fertilized greens and poop from resident Canada geese. (Fortunately, golf courses are surprisingly good at keeping nutrients within the out-of-bounds white stakes.[3]) Yet, the pond never stinks. In the hot months of summer, the pond is aerated by a fountain large enough to churn up the water and thwart disagreeable smells from disturbing the golfers. It's enough to stop the formation of a mini-dead zone.

That same simple idea has been applied on a larger scale to a few water bodies in the United States and Europe. There has been talk of putting a string of big aerators in the main stem of the Chesapeake Bay; a more modest system, consisting of 830 meters of aeration pipes and 138 diffusers, has been in operation since 1988 in a small tributary of the bay, Rock Creek.[4] An aeration system has also been used in the Charles River, which separates Cambridge and Boston, and San Diego recently put in a 130,000 pound

Speece cone, basically a giant bubbler, to aerate one of its reservoirs, Lake Hodges.[5] Like for the Chesapeake Bay, a huge aeration system has been proposed for the Baltic Sea. The two aeration projects actually installed in Baltic coastal waters have yielded mixed results.[6] In both projects, well-oxygenated surface water was pumped down into a hypoxic bottom layer. When that was tried in Sandöfjärden, Finland, the bottom waters expanded, lost dissolved oxygen, and turned anoxic, not just hypoxic. The benefit of the surface water oxygen pumped to the bottom was outweighed by greater oxygen use when microbes were stimulated by the warm surface water.

More impressive and effective is the aeration system put in place recently in Cardiff Bay, Wales.[7] The Cardiff Bay Barrage Act of 1993 stipulates that the bay should have at least 5 milligrams per liter of oxygen, which is much lower than 100 percent saturation but higher than the 2 milligrams per liter that defines hypoxia. When it became clear that the bay would too frequently drop below 5 milligrams per liter, an elaborate aeration system was put into place. It started operations in May 2018. The system consists of 28 kilometers of pipes connecting 6000 diffusers that pump compressed air into the bay's bottom water. While delivering oxygen, the aeration system also increases circulation and mixing, disrupting the stratification that sets up on hot, windless days, those most likely to see hypoxia. Stratification is also weakened when salty bottom water is sucked back out to sea via a saltwater shaft. The system has helped Cardiff Bay maintain oxygen levels above the 5 milligrams per liter limit.

When even 6000 air diffusers aren't sufficient to reach the 5 milligram per liter target, Cardiff Bay turns to an oxygenating barge to pump oxygen gas into low-oxygen waters. Similar barges have been used in Shanghai and Hong Kong.[8] London has two of these barges, the *Thames Bubbler* and its twin the *Thames Vitality*. When in action, the barges transform liquid oxygen into a gas and then inject it into a stream of Thames water, which is then returned to the river. The barges are only summoned in the summer when hot days stimulate oxygen depletion, and heavy rains overwhelm the wastewater-treatment plants, leading to the release of untreated sewage. The number of days the *Thames Bubbler* and *Thames Vitality* are called into action vary greatly from year to year, depending on the weather.[9] In 2016, the barges didn't go out at all, whereas in 2013 they sailed 16 times over 84 days. Instead of barges, two "supersaturation (SSS) oxygenation" plants have been built on the banks of the Swan River estuary in Perth, Australia.[10] Using liquid oxygen like the barges, these plants eliminate hypoxia and restore dissolved oxygen levels for several kilometers upstream and downstream of the plants.

But aeration systems, oxygenation barges, and even supersaturation oxygenation plants are not the way to revive most dead zones. Addition of aluminum to bind up sediment phosphorus has helped some small lakes and semi-enclosed bays,[11] but it doesn't get at the root of the problem, and it isn't feasible for larger waterbodies like the Baltic Sea or Lake Erie. The most effective solution for water near cities is wastewater treatment. Even a fleet of oxygenating barges would not be enough to prevent a return of the Great Stink days to the River Thames if London did not adequately treat its sewage. Wastewater treatment is the main reason why 70 water bodies are no longer dead zones.[12] But that leaves nearly 900 around the world still on the list. Several of those locations in poor countries could be solved by the installation of wastewater-treatment plants, and even in rich countries, untreated sewage is released too often when heavy rain overwhelms the combined sewage-storm overflow system common in old cities, like London. If the city separated its sanitary and storm sewers, the *Thames Bubbler* and *Thames Vitality* would see much less action. This fix is needed elsewhere in Europe and in the United States. In these rich countries, water quality could be further improved if more treatment plants were upgraded to include tertiary treatment to remove nitrogen and phosphorus nutrients.

The treatment plants came about because regulations like the Clean Water Act of 1972 in the United States forced and helped cities financially to build them. Two other regulations have helped some dead zones to a limited extent. As mentioned in Chapter 7, the banning of phosphorus-rich detergents reduced the input of phosphorus nutrients and contributed to the partial recovery of the Baltic Sea and lakes around the world, including Lake Erie. The second set of regulations was written up without water in mind. The US Clean Air Act, first passed in 1963 and last amended in 1990, targeted air pollution, but it also helped aquatic environments by reducing "atmospheric deposition." With fewer pollutants emitted by cars, factories, and power plants, less nitrogen rains down or is deposited into lakes, estuaries, and coastal waters. Atmospheric deposition is one reason why nitrogen concentrations rose in the Chesapeake Bay from 1950 to 1989, and air pollution controls have contributed to the decrease in those concentrations ever since.[13]

In spite of treatment plants, phosphorus detergent bans, and air pollution control, hypoxia is still a problem in the Chesapeake and the Baltic, not to mention the Gulf of Mexico and Lake Erie and the nearly 1000 other dead zones around the world. These dead zones continue to exist today because of excessive nutrients from agriculture, mostly from fertilizers. To revive many dead zones, we need to do something about the fertilizers.

Nutrient Pollution and a Lawsuit in Iowa

Reports about the fertilizer problem in Iowa, the heart of the US Corn Belt, often begin with a comment about the Gulf dead zone, a thousand miles away. But the first casualty of excessive fertilizer use is in Iowa. Over 50 percent of Iowa's rivers and streams are below federal water quality standards and are too polluted for swimming or fishing.[14] Rivers like the Iowa and the Cedar stink.[15] Other rivers like the Raccoon have too much nitrate, as do many of the 150,000 private wells used for drinking water in the state. Nitrate causes methemoglobinemia, better known as blue-baby syndrome; high nitrate levels have been associated with increased risks of bladder and ovarian cancers in Iowa.[16] Even low levels of nitrate may lead to the formation of N-nitrosoamines known to be carcinogenic. Although some studies say the health risks have been overblown,[17] the United States and many other countries have laws stipulating that nitrate in drinking water should be lower than 10 milligrams per liter or 10 ppm, which is at least 10 times higher than natural levels. These health risks are among the reasons why the cost of nitrogen pollution may be as much as $800 billion per year worldwide.[18] The price tag of phosphorus pollution in the United States alone has been estimated to be $2.2 billion per year.[19]

The cost of abating nutrient pollution was the reason why the CEO of the Des Moines Water Works, Bill Stowe, filed a lawsuit against three counties in northwest Iowa. I first came across Stowe in Art Cullen's book *Storm Lake*, which he wrote after winning a Pulitzer Prize in 2017 for editorials critical of industrial agriculture.[20] The editorials appeared in *The Storm Lake Times*, a twice-weekly newspaper published in the northwest Iowa town of the same name, population of about 11,000, where industrial agriculture rules. Cullen tells the story of sitting down next to Stowe at the annual meeting of the Iowa Environmental Council in 2013. Stowe told him that he was going to sue Cullen's county and a couple others in order to stop the counties' nitrate from polluting the city's water and to get them to help pay the city's water treatment costs. The counties, Buena Vista, Sac, and Calhoun, are in the watershed of the Raccoon River, the primary source of drinking water for the Des Moines metropolitan area. The watershed has many heavily fertilized cornfields lined with drainage ditches and underground tiles. As for land formally covered by the Great Black Swamp in Ohio (Chapter 7), the drainage system makes row crop farming possible in this part of Iowa, while also hurrying nutrients from farms to neighboring streams and rivers. Although farmers lose money when fertilizer nutrients leach away, they would lose even more if fields weren't sufficiently drained. Corn in particular doesn't like "wet feet."

Stowe's lawsuit targeted 13 drainage districts overseen by the three counties' elected supervisors. These and 3000 other Iowa drainage districts were created in the early 1900s to lower the water table and "improve" the land for agriculture. The lawsuit contended that the districts are point sources, like pipes coming from a wastewater-treatment plant, and that the drainage districts did not have the proper permits as required by the Clean Water Act of 1972 and related Iowa laws. The lawsuit alleged that the districts were illegally polluting Des Moines's water supply with nitrate, along with pesticides, herbicides, and other chemicals used in agriculture.[21]

In addition to Raccoon River, the city can turn to the Des Moines River, but that river has its own problems: high levels of phosphorus as well as nitrate and possible toxic cyanobacterial blooms.[22] The city cannot risk going through what Toledo experienced in 2014 when a bloom shut off its water supply for two days (Chapter 7). When nitrate levels are too high, Des Moines needs to turn on its massive multimillion-dollar nitrate-removal system, the largest in the world, costing $7000 per day, which added up to $1.2 million in 2015.[23] There are additional costs in dealing with the wastes generated by the system. Recouping those expenses and stopping the nitrate pollution were the reasons behind Bill Stowe's lawsuit.

Stowe was perhaps uniquely suited to lead the Water Works when it submitted the lawsuit.[24] Probably more important than his master's degree in engineering, another in labor relations, and a law degree from Loyola University, Stowe had a "calm demeanor" and spoke with a "tranquil authority" while representing the city during crises or in explaining nitrate pollution to a farmer. Stowe's image was further enriched by his flowing white hair and his resemblance to Russell Crowe, attracting legions of fans who even set up The Bill Stowe Fan Page on Facebook. Stowe loved music, especially the blues, and taking rides on his Harley. He regularly helped with Saturday Mass at St. Anthony Catholic Church near where he and his wife and son lived.

The lawsuit eventually went before the Iowa Supreme Court on January 11, 2016.[25] Six days later the Supreme Count announced its verdict: it decided for the drainage districts and against Stowe and the Des Moines Water Works. According to the Court, Iowa has long given drainage districts immunity because the service they provide—turning swamps and marsh into farmland—was invaluable to the citizens of the state. The Water Works then went to the US District Court with its original argument that the drainage districts were point sources of nitrate pollution. On March 17, 2017, the US District Court dismissed the lawsuit on procedural grounds, citing the Iowa Supreme Court decision and the drainage districts' immunity. The District Court concluded that, while the Des Moines water supply may be polluted by nitrate, the drainage districts lacked the authority to do anything about the problem. Crucially, the court did not weigh in on whether a drainage system with its tiles, pipes, and ditches was a point source. In April, the Des Moines Water Works decided not to appeal the decision and ended its two-year legal battle.

Two years and a month after the lawsuit's defeat, Stowe died of pancreatic cancer. He was 60 years old.

How Farmers Reduce Nutrient Pollution

Let's imagine a more pleasant end to the story. Stowe beats the cancer and continues to ride his Harley in the countryside outside of Des Moines. Along with Cullen and *The Storm Lake Times*, he continues to fight for clean water and against pollution from agribusinesses. Let's also imagine that Stowe and the Des Moines Water Works win their lawsuit against the drainage districts. The courts agree that the drainage systems are point sources of nitrate pollution and that a farm should be regulated like a factory that wishes to dispose of its wastes into a waterbody. Now what? How could drainage districts stop or at least reduce the nitrate and other nutrients leaching from fields and barns? Perhaps one solution would be to install mini-versions of Des Moines's multimillion-dollar nitrate-removal system at each farm. Maybe every farm should have its own treatment plant. Needless to say, that's not going to happen. Of the over 87,000 farms in Iowa, few would be able to afford a treatment plant like Des Moines's even with government support. Even if the lawsuit had been successful, it wouldn't have had an impact on cropland without drainage ditches and underground tiles. Cropland with a drainage system makes up 53 percent of farmland in Iowa, the highest in the nation,[26] but many farms don't need tile drainage.

Fortunately, some farms already have natural treatment plants: buffer zones or strips of natural vegetation that filter out nutrients, other chemicals, and sediments before they pollute neighboring waterbodies (Fig. 10.1). Although often called riparian buffer zones, they are used to border streams, lakes, and drainage ditches as well as rivers. Buffer zones are bypassed by tile drainage, but they help minimize runoff from untiled cropland. Buffer zones come in different sizes and are populated by different plants. The most frequently debated aspect is the minimum width needed for sufficient removal of nutrients.[27] The estimated width for 100 percent effectiveness ranges from 10 meters (about 33 feet) to 30 meters (nearly 100 feet). A meta-analysis published in 2019 of 46 studies surprisingly concluded size didn't matter; the scientists who did the meta-analysis couldn't find a significant relationship between nitrate removal and buffer zone width, which ranged from 10 meters to 200 meters, probably because the variability caused by other factors obscured any effect of buffer zone width. In any case, Iowa's recommendation of 33 feet seems common. The agriculture extension service of Iowa State University says the best buffers have rows of trees and shrubs next to the stream and then perennial grasses on the outside next to the field.[28] According to the meta-analysis, a buffer zone with both trees and grass removes over 90 percent of nitrate, while a grass-only buffer is about 60 percent effective. Even a grass-only buffer may be sufficient to lower nitrate levels

Figure 10.1 Buffer zone with trees and grasses, designed to soak up nutrients before they reach the stream or ditch.

enough to meet the 10 ppm limit for drinking water.[29] Other work has found that removal of phosphorus is roughly the same as for nitrogen. Buffer zones have other benefits besides removing nutrients and sediment. Depending on the vegetation, the wood can be harvested, and they provide habitat for wildlife on land and shade for stream biota.

Farmers are doing a couple of other things that stop nutrients before they even reach a buffer zone. Some of these conservation strategies were summarized in a study led by Don Scavia that explored how to reduce phosphorus inputs into Lake Erie.[30] We first met Don in Chapter 5 when he was hashing out targets for the dead zone in the Gulf of Mexico. Soon after that work, he moved north to the University of Michigan where he turned his attention to Lake Erie's algal bloom and hypoxia problems. Don and his colleagues concluded that several combinations of conservation strategies could be effective, if more widely used.

One combination Don highlighted consisted of buffer zones, subsurface injection of fertilizer, and winter cover crops. We've talked about buffers enough already. As implied by the name, subsurface injection puts fertilizer below the soil surface and surrounds the nutrients with soil, making them less susceptible to runoff loss during a rainfall. Cover crops are not meant to be harvested but rather to cover soils that otherwise would be bare in winter. They increase soil organic matter and water retention, reduce soil erosion, and suppress weeds and diseases. Most relevant here, cover crops soak up nutrients that otherwise would leach away from a bare field. For that reason, these plants are also called "catch crops." Don's study focused on cereal rye because it is cheap and grows fast in the cool weather of Ohio. Other cover crops are also used. Legume cover crops like alfalfa have been called "green manure" because they have symbiotic nitrogen-fixing bacteria that add useable nitrogen to soils. While cycling past farms in Delaware near where I live, I've seen fields strewn with what I later learned were daikon radishes, which are used as a cover crop in this part of the country. In addition to taking up nutrients before they leach away, the radish's large taproot breaks up compacted soils.

Don Scavia's paper doesn't say a lot about crop rotation, but other studies have found that rotation among several row crops can reduce nutrient loss.[31] Compared to the common practice of rotating between corn and soybean, rotating with oat, clover, and alfalfa reduces total nitrogen runoff by 39 percent and phosphorus runoff by 30 percent without affecting yields and profits. Although the Scavia-led study was about phosphorus and Lake Erie, the nutrient-reduction strategies, if not the exact combinations and recommendations, are applicable to lowering nitrogen nutrients and to solving hypoxia problems elsewhere in the United States and the world.

The Scavia study doesn't explicitly say anything about another strategy that could lower nutrient runoff from agriculture, that being improvements in nutrient use efficiency (NUE), which is the amount of a nutrient taken up by a crop relative to the amount added as fertilizer. A higher NUE means more nutrients are absorbed by the crop and less is left behind to leach away to nearby waterbodies. The efficiency of crops using nitrogen and phosphorus has improved over the last 20 years because of advances in crop breeding, more testing of soils and crop tissue, and application of the "4Rs":[32] the fertilizer applied at the right rate, in the right place, at the right time, and in the right form of the nutrient (nitrate vs. ammonia vs. urea, for example). Today the average American farmer is not getting at least one of the Rs quite right: they purposely add more nitrogen fertilizer than the recommended rate, as a kind of insurance.[33] If the weather cooperates, the additional fertilizer will help achieve higher yields. Smart use of fertilizer is part of "precision agriculture," which has been touted as a "painless way" to reduce dead zones. While that remains to be seen, higher NUEs would certainly help. One study exploring how to shrink the Gulf of Mexico dead zone suggests that the nitrogen NUE for corn and other row crops needs to increase by 43 percent. Perhaps it can go higher, but the nitrogen NUE in the United States is already among the highest in the world.[34] There is much more room for improvement in China where the NUE is 0.25 versus 0.68 in the United States and Canada.

So, it seems that farmers know how to lower nutrient pollution, and in theory, we have the solutions to the dead-zone problem. Farmers already are using many of the conservation strategies we just discussed. The Scavia-led study mentioned that at least one of the strategies is being used on 99 percent of the farms in the watershed of western Lake Erie. But the lake still has hypoxia and algal bloom problems, and there is still a dead zone in the Gulf of Mexico and a litany of eutrophication-caused problems in waterways in the US Midwest and the rest of the country. And of course, the Baltic Sea and the Black Sea dead zones haven't gone away either.

The conservation strategies as now being implemented haven't been enough in the United States, in part because the country has relied mostly on carrots rather than sticks to motivate farmers to reduce nutrient pollution. The carrots—the financial incentives to encourage conservation—have been small compared to what farmers get from growing corn and other row crops.

At the federal level, money for buffer zones and similar conservation efforts comes from the Conservation Reserve Program (CRP), which is administered by the US Department of Agriculture.[35] Started in 1985, CRP pays out each year a rent to farmers who "agree to remove environmentally sensitive land from agricultural production and plant species that will improve environmental health and quality." The contracts last 10 to 15 years. The Conservation Reserve Enhancement Program, which is part of CRP, draws money from both federal and state sources. The CRP alone gave out $1.85 billion in 2019, or on average $137 per acre in 12 states making up nearly all of the critical conservation area of the Mississippi River basin.[36] (I did not include the sliver of South Dakota for this and the analysis to follow.) The average for Iowa was $222. To a city slicker, that may sound like a lot of money, but it is far less than the $700 a farmer could get from an acre planted in corn, assuming a yield of 196 bushels per acre and each bushel fetching $3.14 (2018 estimates). Perhaps not surprisingly, some farmers break their CRP contract and pay the fine, knowing they can make more money by planting corn. These numbers are why the Des Moines Water Works lost their lawsuit and why Iowa laws have favored drainage districts that have turned wetlands into cropland. They are also the reason why the Great Black Swamp in Ohio was drained and why bringing part of it back to stop nutrients flowing into Lake Erie[37] is unlikely.

Chris Jones, who had counted up livestock and people in Iowa (Chapter 5), made another comparison about the scale of Iowa's conservation efforts.[38] He pointed out that the state of Iowa provides about $14 million per year to farmers for conservation efforts to lower nitrate levels, but that effort is negated by the money, $70 million in 2016 alone, farmers spend on tile drainage. The bottom line is that the carrots from the CRP program are not enough when so much money can be made from corn.

Here's still another comparison to put the conservation efforts in perspective. Although the billions of dollars already spent on conservation efforts isn't chump change, it's small compared to the money that built wastewater-treatment plants and that went to stop other major point sources of nutrients. As mentioned in Chapter 1, the price tag for treatment plants in the United States alone has been about $810 billion. Since passage of the Clean Water Act in 1972, total cleanup efforts by municipalities clock in at $1.04 trillion, and industries have spent another $571 billion (Fig. 10.2).[39] The money spent on controlling all point sources is about eight times more than the $193 billion spent to stop nonpoint sources. The allocation of money for controlling point versus nonpoint sources is the opposite of the relative importance of those

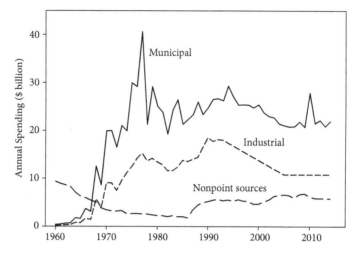

Figure 10.2 Money spent in the United States to combat pollution from point sources (municipal and industrial) and from agriculture and other nonpoint sources.

two sources in contributing to nutrient pollution. The input of nutrients from nonpoint sources—agriculture—is several times higher than the input from point sources to waterbodies in the United States and Europe.

One reason I give these numbers is to put the price tag for reviving a dead zone in perspective and to suggest that money is not the only reason why voluntary conservation efforts haven't been enough. One study estimated that reaching the Gulf of Mexico Hypoxia Task Force target would cost $2.6 billion per year.[40] Before I gathered the numbers just mentioned, I thought $2.6 billion was huge and an insurmountable barrier to solving the Gulf's deadzone problem. It is a lot of money but not compared to what has been spent on conservation efforts so far. Don Scavia calculated that over $30 billion has supported those efforts since 1995.[41] I figure that farmers in the Mississippi critical conservation area states have received $14.9 billion from CRP alone.[42] Not all that money goes to fighting nutrient pollution, and I'm sure some conservation efforts attacking the hypoxia problem could use more resources. Farmers could be educated about the benefits of cover crops and given help with subsurface injection of fertilizers and testing soil for nutrient levels; studies have shown that farmers who do soil testing use less fertilizer.[43] But I suspect any new conservation efforts will not require vast amounts of new financial support. The lack of money isn't the main reason why we still have eutrophication problems and dead zones in the United States. I don't think money alone is the solution.

Doing More than "Paying the Polluters"

It is clear that the carrots-only approach hasn't worked. Paying farmers to stop nutrient pollution hasn't been enough to alleviate eutrophication problems in local waterways and to revive dead zones like those in the Gulf of Mexico and Chesapeake Bay. What's missing is the willingness of government to take a more active role in getting something done about the problem.

The US federal government has mostly left regulations of nonpoint sources of pollution up to the states, and with few exceptions, the states in turn have made local governments responsible for watching over local waterways. The result has been an ineffective patchwork of regulations and recommendations. Buffer zones are a good example. Most of the states in the Mississippi River basin have only recommendations, not laws, for buffer zones, whereas statewide regulations mandating buffer zones are somewhat stronger in the Chesapeake Bay watershed. Maryland and Virginia require 100-foot buffers (about 30 meters) for land bordering the Chesapeake Bay, and counties within each state can impose further restrictions on streams outside of the Chesapeake Bay Critical Area (as it's called in Maryland)[44] or the Resource Protection Area (Virginia).[45] However, Pennsylvania, the state that sends the most nutrients to the Chesapeake via the Susquehanna River, doesn't have any statewide regulations, and has relied on county governments to oversee conservation efforts. Even where buffer zones are required, there is often no mechanism to ensure they are constructed and maintained properly. Newly planted saplings need attention if they are to survive and grow to trees large enough to trap nutrients.

Another difference between the US Midwest and the Chesapeake Bay watershed states is how they handle manure, specifically what to do with it in winter when its nutrients are more likely to leach into nearby waterbodies.[46] The regulations again vary from state to state. The Corn Belt states have fairly lax regulations and allow some application to fields in winter. In Iowa, for example, although farmers cannot spread manure on frozen fields, they can ask the state's Department of Natural Resources for a waiver in case of emergencies. Among states in the Chesapeake Bay watershed, Pennsylvania's regulations are only a bit stricter than seen in the Corn Belt, while Maryland has some of the strictest in the country. The state prohibits applications between November 2 and February 28 in eastern Maryland and November 16 and February 28 on the west side of the state. Four northern European countries—Denmark, Norway, Sweden, and Finland—also prohibit winter applications.

Regions with stricter rules about manure generally have cleaner waters. Denmark has some of the strictest regulations and has made the most progress, which will be discussed later. At the other end of the spectrum, China had no regulations until recently and has had serious eutrophication problems. Worse than having few laws about winter applications, disposal of manure has been too lightly regulated year-round; as much as half of the manure produced in China may be dumped into rivers.[47] It is still too early to say whether the regulations implemented in 2017 to curtail manure pollution are effective. As mentioned in Chapter 5, manure is thought to be a large source of pollution in China, fouling the country's air as well as its waterways.

Alongside buffer zone and manure regulations, another approach would be to focus on results: reducing nutrient levels. Scientists on the Hypoxia Task Force argued for this in the late 1990s as the way forward to shrink the Gulf of Mexico dead zone. In Chapter 5, I mentioned that the Task Force considered a total maximum daily load (TMDL) for nitrogen nutrients, but that was argued down by Task Force members from the Corn Belt states. Seeing the lack of progress in the Gulf and Lake Erie, environmental groups sued the US EPA in the 2000s to take a more active role in stopping nutrient pollution. The lawsuits were unsuccessful. There has been some progress in reducing nutrient concentrations in some regions of the Mississippi River basin, but that progress has been offset by rising concentrations in other regions.[48] The net result is that nutrient inputs into the Gulf have not decreased, and the dead zone is not shrinking. As I mentioned in Chapter 7, the Task Force pushed back to 2035 the goal of reducing the dead zone to 5000 square kilometers and set an interim goal to reduce nitrogen and phosphorus nutrients by 20 percent before 2025. Unless things change drastically, I doubt even that modest goal will be met.

The EPA and state governments have a different relationship in the Chesapeake Bay region. In 2010, the EPA stepped in to set up a TMDL plan at the request of the six Chesapeake watershed states and the District of Columbia when they saw the Bay's environmental health decline.[49] The Clean Water Act prohibits the EPA from enforcing TMDLs for nutrients, and the states are still responsible for coming up with plans for reducing nutrient levels. But the EPA can penalize a delinquent state for not meeting interim goals. This active role of EPA has been credited for some of the success in reducing nutrient inputs and improving the environmental health of the Bay,[50] although skeptics would say those improvements were mostly due to better wastewater treatment and air pollution controls.

Much less has been done about solving the harder problem of reducing nutrients coming from agriculture. Each of the six states in the Chesapeake Bay watershed has a plan in place to reduce nutrient pollution by about 40 percent (relative to 1985) by 2025. If they are to meet the 2025 deadline, these states somehow need to do what they haven't come close to doing so far: stop farms from polluting the states' waterbodies. To reach the deadline, nutrients from agriculture have to decrease tremendously. It's clear that's not going to happen. Less clear is how EPA will respond. It doesn't have many legal tools to force a state into action, but any EPA response could signal a new chapter in the fight against nutrient pollution in the United States.

The United States could learn much from what Denmark has done to reduce nutrient pollution. Denmark was one of the first countries in Europe to recognize that its coastal waters, including regions in the North Sea to the west and in the Baltic to the east, were suffering from eutrophication problems. So, starting in 1987, long before the Nitrate Initiative was imposed by the European Union, Denmark passed its first Action Plan for the Aquatic Environment.[51] The plans have become increasingly aggressive and ambitious over the years since 1987. Most notably, the Danish plans have many regulations to reduce nutrients from nonpoint sources; agriculture covers about 60 percent of the country's land and uses a lot of commercial fertilizer and manure. Aimed at reducing those nutrient sources, the plans mention winter cover crops, catch crops, and creating wetlands and forested buffers. The amounts and even types of manure are regulated: "pig slurry: 75 percent, cattle slurry: 70 percent, deep litter: 45 percent, other types: 65 percent." Danish farmers have to maintain a nitrogen budget. Although at first they could use only enough fertilizer to yield 90 percent of maximum profits, in 2016, the law was loosened so that farmers now can add enough fertilizer to achieve maximum profits but only if they take other, compensatory measures to reduce nutrient pollution.[52] The country has been much more aggressive than other Baltic watershed countries in combating nutrient pollution.

Denmark's Action Plan has worked, although there have been some disappointments.[53] As seen elsewhere, better sewage treatment plants had the biggest, most immediate effect, reducing nitrogen inputs by 75 percent and phosphorus inputs by more than 90 percent since the mid-1980s. But unlike in the United States, with its laissez faire approach, nitrogen inputs from agriculture also have decreased by over 40 percent, although phosphorus inputs haven't declined beyond what the treatment plants had already achieved. Lower inputs have translated into lower nutrient levels in coastal waters; total nitrogen and phosphorus concentrations have been cut in half. The rest of the story is not as positive. Although algal biomass has declined by more than 50 percent, and seagrass meadows have expanded into deeper waters as the water became clearer, dissolved oxygen in bottom waters has not increased.

Danish scientists think that the decline in nutrient levels was offset by the other main ingredient producing hypoxic bottom waters: stratification. From 1980 to 2015, stratification of Danish coastal waters was strengthened by warming temperatures, and it was broken down less frequently by wind because wind speeds have been slower since 1990. Also, warmer water holds less oxygen and stimulates more use of the gas by microbes. In short, reducing nutrients wasn't enough to counteract climate change.

That's the fear of ecologists and environmentalists working on reviving dead zones. Because of climate change, nutrient inputs probably need to be reduced even more than now projected. As seen in the open oceans discussed in Chapter 9 and just mentioned for Danish coastal waters, warmer waters alone would be enough to worsen dead zones.[54] But just about every facet of climate change, from sea level rise to the frequency of intense storms, will affect the initiation, the area and volume, and the duration of hypoxic waters around the world.

The few actions taken by governments to combat nutrient pollution are partially offset by other governmental policies. While US and European governments have spent a lot of money to control nutrient pollution, other governmental policies lead to more nutrient pollution. One example is agricultural subsidies. The European Union gives its farmers about $65 billion per year or nearly 40 percent of its total budget. As described in a recent *New York Times* article,[55] there is a link between the amount of the subsidies given to a region within the EU and the level of nitrate in streams and rivers of that region. The article finds fault with several countries, but it highlights Poland's contribution to eutrophication and algal blooms in the Baltic Sea. That country contributes the most nitrogen and phosphorus to the Baltic in part because it has more farmland than other countries in the Baltic watershed.[56] Those farms are supported by generous EU subsidies; Poland receives more than most other EU countries, except for France, Spain, Germany, and

Italy. But Poland is not alone, and there are more EU countries like it than like Denmark when it comes to controlling nutrient pollution. Given current EU subsidies and conservation efforts, the prognosis for many regions of the Baltic Sea is grim.[57]

Direct agricultural subsidies are smaller in the United States, but American corn farmers receive a huge indirect subsidy via the Renewable Fuel Standard.[58] Passed in 2005, the law requires that gasoline sold in the United States must contain 10 percent ethanol. The ethanol is renewable (it's a "biofuel") because it can be made by fermenting sugars from plants. Although the EPA was given the authority to require ethanol to be made with crops like switchgrass and sugar cane, the most abundant source turned out to be corn. Now about 45 percent of the corn harvested in the United States goes to making ethanol. Yes, you read it correctly: 45 percent. Some of the leftovers from ethanol production, equivalent to a third of the initial corn, is used as animal feed. Even so, a lot of corn ends up in American gas tanks.

To fill those tanks, corn production ramped up in the Corn Belt after en-actment of the Renewable Fuel Standard. From 2007 to 2012, 3.2 million acres of unused land and one million acres of environmentally sensitive land once protected by the CRP were converted to cornfields.[59] Farmers increas-ingly planted corn "fence to fence"—no buffer zones, and "corn on corn"—one year of corn followed by another without rotating to another crop. The ethanol-fueled increase in corn production, of course, has led to more nutrients ending up eventually in the Gulf of Mexico, as much as 18 percent for nitrate and 12 percent for phosphorus.[60] The nutrient inputs due to corn ethanol are a big fraction, perhaps as much as a half, of the amount nutrients need to decline in order to shrink the Gulf dead zone to the targeted size.

A political scientist needs to take the story from here in order to explore how governments could do a better job at stopping nutrient pollution. Suffice it to say, something needs to change. One summary of the needed changes is by Don Boesch.[61] He pointed out that our current system, built on economic incentives and subsidies, is to "pay the polluter." Boesch thinks what we need more of is "the polluter pays." As with all aphorisms, this one is too simplistic. I don't think a Corn Belt farmer is the same as the Exxon Mobile corporation, and nitrate is a different type of pollutant than oil. More importantly, the en-tire industrial agriculture business complex, of which the farmer is only one cog, needs to be scrutinized. But Boesch's aphorism captures a truth. It points out how the mindset of governments needs to change if eutrophication and dead-zone problems are to be solved.

Doing Your Part

Some of the people working to revive dead zones are obvious: those operating wastewater-treatment plants and aeration systems, and manning the oxygenation barges and plants. Of course, there are the farmers and others in agriculture trying to reduce fertilizer use and runoff. It may not be clear what the rest of us can do, other than to pressure our governments to address the problem more vigorously. But we can do something by changing what most of us do three times (or more) a day: eat. Each of us can effect change in agriculture by how much and what we eat. A better diet would not only improve our health, it would also help the environment, including the climate. Agriculture and the food industry release nearly a third of all greenhouse gases emitted by anthropogenic sources.[62] Our diet is also connected to fertilizer use, nutrients in waterways, and the dead-zone problem.

The first step is to eat less. The well-known health problems of overeating and obesity include cardiovascular diseases, type 2 diabetes, and some cancers. Less appreciated is the impact on the health of the environment. In a recent study that looked at the environmental impact of overeating and obesity, the authors calculated that worldwide, people eat 140.7 million tons of food more than is necessary for a healthy diet.[63] Most of this excessive food is consumed in the European Union (39.2 million tons), followed closely by North America (32.5 million tons). All those tons of excessive food translate into a hefty load on the environment. Unfortunately, no one has examined how much fertilizer is needed to produce the 140.7 million tons of excessive food, but it must be substantial. About 67 percent of all Americans are overweight and 36 percent are obese.[64] People in other rich countries are not far behind. Those percentages suggest that cutting back on what we eat would also significantly cut back on the nutrients feeding dead zones.

Along with eating less, we need to change the types of food we eat. The "we" here includes the average North American, many Europeans, and the increasing number of Asians who are adopting a Western, beefy diet. The take-home message is the same as just mentioned: eating less meat improves our health and the health of the environment.[65] Red meat stands out as being particularly harmful. Eating too much red meat is bad for our health and raising cows is bad for the environment. To compare the climate impacts of cows and cars, scientists put the different greenhouse gases produced by both in terms of "carbon dioxide equivalents." Methane, the main ingredient in natural gas, is over 20 times more effective than carbon dioxide in trapping heat, so one methane molecule is equal to over 20 carbon dioxide equivalents. The large amount of methane belched and farted out by cows is one reason why producing red meat emits over 40 times more carbon dioxide equivalents than does producing an equivalent serving of vegetables. The damage in terms of nutrient pollution is even greater: over 50 times more phosphorus is released. Poultry is better for us and the biosphere around us; producing a serving of chicken or turkey releases a third less carbon dioxide equivalents and a third less nutrients than red meat production does. But best for us and the environment would be a diet with less meat of all kinds.

There are other, looser connections between our food, our health, and dead zones. Drinking fewer sugary beverages would lower our caloric intake and ease obesity-related health problems, but it wouldn't do much for the environment, at least in terms of fertilizer use. The sweet taste in these beverages made in the United States is thanks to high fructose corn syrup. The fraction of the total corn harvest devoted to making this syrup is small, less than 5 percent,[66] so cutting soft drinks out of the American diet probably would not reduce fertilizer use by much. The sweetener added in other countries, sucrose, comes from sugar cane and beets, which use much less fertilizer than required to grow corn.

By contrast, cutting down on food waste could reduce fertilizer use substantially. In the United States, as much as half of all food is thrown away,[67] whereas food waste is about 18 percent of total production in Europe.[68] Eliminating that waste would not reduce fertilizer use as much as what those percentages suggest because most of the wasted food are cereals and vegetables with low nitrogen and phosphorus content. Even so, the savings in fertilizer could still be substantial, estimated to be about 16 percent in Europe. The human health connection is more complicated. The health of many people would be improved if the food now wasted were used to help alleviate food insecurity and malnutrition, but nutrient pollution would continue without changes in agriculture. We need to figure out how to feed the world's growing population while also reducing the environmental costs of agriculture.

Of course, we cannot eat our way to solving climate change and dead-zone problems. There are larger sources of greenhouse gases than agriculture, and there are many steps between what's on our dinner plate and the nutrients leaching from fields and barns into waterways. Climate change will be abated only when we stop driving gas-guzzling SUVs and turn down the thermostat in winter and turn it up in summer. Dead zones will be revived only when we reduce fertilizer use and enact and enforce conservation measures to stop nutrient pollution. The point is, the problems of climate change, dead zones, and human health are connected. The studies about these connections argue that progress in solving one will go toward solving the others. The hope is that if people are made aware of, become concerned about, and act on one problem, then they will do the same for the others. One way or the other, our diet, climate change, and dead zones are intertwined. Eating fewer steaks and hamburgers doesn't sound like much, but it could go a long way to solving some of the most important environmental problems facing society today.

Will we be able to revive dead zones? At the end of my conversations with the experts, I would ask if they were optimistic about solving the problem. They would pause to gather their thoughts and then give a guarded answer, some more optimistic than others, about one dead zone more so than another. At first, my answer to the question was pessimistic. It is hard to see much progress today in reducing nutrients and raising oxygen levels in the Baltic Sea and the Gulf of Mexico. But then I realized that no one who lived through the Great Stink of 1858 would have guessed that salmon would ever again swim under London Bridge, and few in wartime Philadelphia would have predicted that residents and tourists would ever want to kayak on the Delaware River. Dissolved oxygen has returned to a handful of other water bodies as well. The forces working against improvements elsewhere seem formidable now, yet that was also the case in the early days of the fight against DDT and phosphorus detergents. The bans against those chemicals are now environmental success stories. Even if they are not perfect harbingers of oxygen's return, the successes achieved so far offer reason to be optimistic about reviving dead zones around the world.

Let's hope the optimists are right.

Acknowledgments

While working on this book, I learned once again that most scientists are generous, supportive, and patient in answering even the oddest question. That was the case with everyone I talked to or emailed about this project. Don Boesch provided stories behind the publications, sent hard-to-get reports, and promptly answered my many emails. Always gracious, Nancy Rabalais thoroughly addressed my questions and commented extensively on several of the chapters. In addition to telling me the backstories of his work in the Gulf of Mexico, Gene Turner sent old data essential for some of my analyses, reminded me of papers I had forgotten, and provided remarks on much of the book. Jacob Carstensen, Daniel Conley, and Don Scavia were great sources of information, and all three reviewed several chapters. Jean Brodeur made many perceptive points, and Tiffany Strasa scrutinized every word, pushing me to clean up muddled paragraphs and to think harder about what I was trying to say. The following people, who are either experts in the field or just concerned citizens, also provided valuable feedback: Mike Best, Denise Breitburg, Frances Chen, Ray Graff, John Hanson, Chris Jones, Jerry Kaufman, Susan Lee, Lisa Levin, Bernard Noeller, Janice Pinto, Mike Roman, and Pam Tootson. Many thanks go out to all of these people. Any mistakes or infelicitous phrases remaining in the book are all mine.

I thank again others mentioned in the book who provided useful information or data. Lillian Wang and Dylan Taillie helped with the maps and some of the schematic diagrams. I couldn't have done the GIS analyses without the help of Chaoqun Lu and Pinki Mondal. My apologies to anyone not mentioned here, an oversight that says more about my problem remembering names than about the value of their contribution or the depth of my gratitude. Finally, I thank my wife, Ana Dittel, for her unfailing support and patient tolerance of my bad golf.

Notes

Prologue

1. "Dead zone" was used earlier than 1985 but not to describe low-oxygen waters like in the Gulf of Mexico. Kirchman, D.L. "The first 'dead zone'" *Limnology and Oceanography Bulletin* (2020) 10.1002/lob.10402.
2. Bane, Nikki, and Peter Eldridge, "Annual Report of the Southeast Area Monitoring and Assessment Program (SEAMAP)" (Ocean Springs, MS: Gulf States Marine Fisheries Commission, 1985).
3. http://www.noaa.gov/media-release/gulf-of-mexico-dead-zone-is-largest-ever-measured, accessed June 5, 2018.
4. Vaquer-Sunyer, Raquel, and Carlos M. Duarte, "Thresholds of Hypoxia for Marine Biodiversity," *Proceedings of the National Academy of Sciences* 105 (2008): 15452–57.
5. King, Richard J., *Ahab's Rolling Sea: A Natural History of Moby-Dick* (Chicago: The University of Chicago Press, 2019), p. 37.
6. Keiser, David A., and Joseph S. Shapiro, "Consequences of the Clean Water Act and the Demand for Water Quality," *The Quarterly Journal of Economics* 134 (2019): 349–96.

Chapter 1

1. Faraday, Michael, "The State of the Thames," *The Times*, July 9, 1855, p. 8.
2. Thompson, J. D., "The Great Stench or the Fool's Argument," *The Yale Journal of Biology and Medicine* 64 (1991): 529–41.
3. https://api.parliament.uk/historic-hansard/commons/1858/jul/15/first-reading#S3V0151P0_18580715_HOC_123, accessed July 1, 2019.
4. Wheeler, Alwyne C., *The Tidal Thames: The History of a River and Its Fishes* (London: Routledge & Kegan Paul, 1979), p. 30.
5. Gostelow, P., S. A. Parsons, and R. M. Stuetz, "Odour Measurements for Sewage Treatment Works," *Water Research* 35 (2001): 579–97.
6. Sedlak, David L., *Water 4.0: The Past, Present, and Future of the World's Most Vital Resource* (New Haven: Yale University Press, 2014).
7. Ackroyd, Peter, *London Under* (London: Chatto & Windus, 2011).
8. Wheeler, *Tidal Thames*.
9. https://www.oldbaileyonline.org/static/Population-history-of-london.jsp#a1815-1860, accessed August 21, 2020.
10. https://en.wikipedia.org/wiki/Locks_and_weirs_on_the_River_Thames#External_links, accessed June 26, 2019.

11. Halliday, Stephen, *The Great Stink of London: Sir Joseph Bazalgette and the Cleansing of the Victorian Metropolis* (Stroud, England: Sutton, 1999).

12. UK House of Commons Papers (1840), vol. 11, page 209, q3452, https://archives.parliament.uk/online-resources/parliamentary-papers/, accessed September 28, 2020.

13. Halliday, *Great Stink of London*, p. 124.

14. Brody, Howard, Michael Russell Rip, Peter Vinten-Johansen, Nigel Paneth, and Stephen Rachman, "Map-Making and Myth-Making in Broad Street: The London Cholera Epidemic, 1854," *The Lancet* 356 (2000): 64–68.

15. Halliday, *The Great Stink of London*.

16. Wheeler, *Tidal Thames*, p. 25.

17. Wood, Leslie B., *The Restoration of the Tidal Thames* (Bristol: Adam Hilger, 1982).

18. Wood, *Restoration of the Tidal Thames*, p. 41.

19. Wood, *Restoration of the Tidal Thames*, p. 41.

20. Anonymous, "Biography of the Week: Mr. W.J. Dibdin, F.I.C., F.C.S., Past President of the Institute of Sanitary Engineers," *The Engineering News* (1908): 127.

21. Wood, *Restoration of the Tidal Thames*.

22. Hamlin, Christopher, "William Dibdin and the Idea of Biological Sewage Treatment," *Technology and Culture* 29 (1988): 189–218.

23. Hamlin, "William Dibdin and the Idea."

24. Wood, *Restoration of the Tidal Thames*.

25. Barry, Wolfe, "Discussion of the Main Drainage of London," *Minutes of the Proceedings of the Institution of Civil Engineers* 129 (1897), p. 112.

26. http://www.guise.me.uk/dibdin/wjdibdin/obituaryinpaper.htm, accessed September 23, 2019.

27. Sedlak, *Water 4.0*.

28. Sedlak, *Water 4.0*, p. 45.

29. Dibdin, W. J., "The Purification of the Thames," *Minutes of the Proceedings of the Institution of Civil Engineers* 129 (1897): 80–111.

30. Dibdin, W. J., "Sewage Sludge and Its Disposal," *Minutes of the Proceedings of the Institution of Civil Engineers* 88 (1887): 155–71.

31. Sedlak, *Water 4.0*.

32. Cutler, David, and Grant Miller, "The Role of Public Health Improvements in Health Advances: The Twentieth-Century United States," *Demography* 42 (2005): 1–22.

33. Cutler and Miller, "Role of Public Health Improvements."

34. Hazen, Allen, *Clean Water and How to Get It* (John Wiley & Sons: New York, 1907), p. 36.

35. Levy, Sharon, *The Marsh Builders: The Fight for Clean Water, Wetlands, and Wildlife* (New York: Oxford University Press, 2018), p. 35.

36. Wood, *Restoration of the Tidal Thames*.

37. Sharp, Jonathan H., "Estuarine Oxygen Dynamics: What Can We Learn About Hypoxia from Long-Time Records in the Delaware Estuary?," *Limnology and Oceanography* 55 (2010): 535–48.

38. Albert, Richard C., "The Historical Context of Water Quality Management for the Delaware Estuary," *Estuaries* 11 (1988): 99–107.

39. Albert, "Historical Context of Water Quality Management."

40. Selby, Earl, and Dorothy Selby, "Clean-up on the Delaware," *Collier*, January 5, 1946.

41. Sedlak, *Water 4.0*.

42. Brosnan, T. M., A. Stoddard, and L. J. Hetling, "Hudson River Sewage Inputs and Impacts: Past and Present," in *The Hudson River Estuary*, edited by Jeffrey S. Levinton and John R. Waldman (Cambridge: Cambridge University Press, 2006), pp. 335–48.

43. Meybeck, M., L. Lestel, C. Carré, G. Bouleau, J. Garnier, and J. M. Mouchel, "Trajectories of River Chemical Quality Issues over the Longue Durée: The Seine River (1900s–2010)," *Environmental Science and Pollution Research* 25 (2018): 23468–84.

44. Soetaert, Karline, Jack J. Middelburg, Carlo Heip, Patrick Meire, Stefan Van Damme, and Tom Maris, "Long-Term Change in Dissolved Inorganic Nutrients in the Heterotrophic Scheldt Estuary (Belgium, the Netherlands)," *Limnology and Oceanography* 51 (2006): 409–23.

45. Leopold, Aldo, *A Sand County Almanac: With Essays on Conservation* (New York: Oxford University Press, 2001), p. 172.

46. Carson, Rachel, *Silent Spring* (Boston: Houghton Mifflin Company: The Riverside Press, 1962).

47. Carson, Rachel, *Under the Sea-Wind: A Naturalist's Picture of Ocean Life* (New York: Oxford University Press, 1952).

48. Sedlak, *Water 4.0*.

49. Wood, *Restoration of the Tidal Thames*.

50. Meybeck et al., "Trajectories of River Chemical Quality Issues."

51. These are 2014 dollars. Keiser, David A., and Joseph S. Shapiro, "Consequences of the Clean Water Act and the Demand for Water Quality," *The Quarterly Journal of Economics* 134 (2019): 349–96.

52. Data from environment.data.gov.uk/water-quality. Dissolved oxygen is higher at high tide and lower in the summer. It also varies with location.

53. Meybeck et al., "Trajectories of River Chemical Quality Issues."

54. Cloern, James E., "Patterns, Pace, and Processes of Water-Quality Variability in a Long-Studied Estuary," *Limnology and Oceanography* 64 (2019): S192–S208.

55. Brosnan et al., "Hudson River Sewage Inputs."

56. Griffiths, Andrew M., Jonathan S. Ellis, Darryl Clifton-Dey, Gonzalo Machado-Schiaffino, Dylan Bright, Eva Garcia-Vazquez, and Jamie R. Stevens, "Restoration Versus Recolonisation: The Origin of Atlantic Salmon (*Salmo Salar* L.) Currently in the River Thames," *Biological Conservation* 144 (2011): 2733–38.

57. Georgiou, Aristos, "NYC Shores Flooded with Whales and Experts Think They Know Why," *Newsweek* (May 29, 2019), https://www.newsweek.com/nyc-shores-flooded-whales-experts-1437868, accessed March 17, 2020.

58. Revkin, Andrew, "14-Foot Fish Spotted in River, Giving Hope to Vanished Giant's Return," *National Geographic* (March 7, 2019), https://www.nationalgeographic.com/environment/2019/03/14-foot-atlantic-sturgeon-fish-hudson-river-endangered-species/#close, accessed March 17, 2020.

59. Arroita, Maite, Arturo Elosegi, and Robert O. Hall Jr., "Twenty Years of Daily Metabolism Show Riverine Recovery Following Sewage Abatement," *Limnology and Oceanography* 64 (2019): S77–S92.

60. Kauffman, Gerald J., Andrew R. Homsey, Andrew C. Belden, and Jessica Rittler Sanchez, "Water Quality Trends in the Delaware River Basin (USA) from 1980 to 2005," *Environmental Monitoring and Assessment* 177 (2011): 193–225.

61. Partnership for the Delaware Estuary, *Technical Report for the Delaware Estuary and Basin 2017*, edited by L. Haaf, S. Demberger, D. Kreeger, and E. Baumbach. Report No. 17-07; Partnership for the Delaware Estuary.

62. Breese, G., "Chapter 6.8—American Shad," in *Technical Report for the Delaware Estuary and Basin 2017*, edited by L. Haaf, S. Demberger, D. Kreeger, and E. Baumbach (Report No. 17-07; Partnership for the Delaware Estuary).

63. Sato, Toshio, Manzoor Qadir, Sadahiro Yamamoto, Tsuneyoshi Endo, and Ahmad Zahoor, "Global, Regional, and Country Level Need for Data on Wastewater Generation, Treatment, and Use," *Agricultural Water Management* 130 (2013): 1–13.

64. Ali, Mohammad, Allyson R. Nelson, Anna Lena Lopez, and David A. Sack, "Updated Global Burden of Cholera in Endemic Countries," *PLOS Neglected Tropical Diseases* 9 (2015): e0003832.

65. Yu, Fengling, Yongqiang Zong, Jeremy M. Lloyd, Guangqing Huang, Melanie J. Leng, Christopher Kendrick, Angela L. Lamb, et al., "Bulk Organic Δ^{13} C and C/N as Indicators for Sediment Sources in the Pearl River Delta and Estuary, Southern China," *Estuarine, Coastal and Shelf Science* 87 (2010): 618–30.

66. Ram, Anirudh, Jiyalal Ram M. Jaiswar, M. A. Rokade, S. Bharti, C. Vishwasrao, and D. Majithiya, "Nutrients, Hypoxia and Mass Fishkill Events in Tapi Estuary, India," *Estuarine, Coastal and Shelf Science* 148 (2014): 48–58.

Chapter 2

1. Email from R. E. Turner to author, August 1, 2016, and interview with author, Baton Rouge, LA, January 31, 2019.

2. Bender, M. E., Donald J. Reish, and C. H. Ward, "Re-Examination of the Offshore Ecology Investigation," in *The Offshore Ecology Investigation: Effects of Oil Drilling and Production in a Coastal Environment*, edited by C. H. Ward, M. E. Bender, and Donald J. Reish (Houston, TX: William Marsh Rice University, 1979), pp. 35–116.

3. Turner, R. E., and R. L. Allen, "Bottom Water Oxygen Concentration in the Mississippi River Delta Bight," *Contributions in Marine Science* 25 (1982): 161–72.

4. Rabalais, Nancy N., Leslie M. Smith, and R. Eugene Turner, "The Deepwater Horizon Oil Spill and Gulf of Mexico Shelf Hypoxia," *Continental Shelf Research* 152 (2018): 98–107.

5. Anonymous, "Oxygen Level Seen Low," *Victoria Advocate*, February 5, 1982, p. 19.

6. Officer, C. B., R. Biggs, J. Taft, L. E. Cronin, M. A. Tyler, and W. R. Boynton, "Chesapeake Bay Anoxia: Origin, Development, and Significance," *Science* 223 (1984): 22–27.

7. Interview of Charles Officer and Trixie Officer by Ronald Doel on November 29, 1995, Niels Bohr Library & Archives, American Institute of Physics, College Park, MD,

www.aip.org/history-programs/niels-bohr-library/oral-histories/6992, accessed April 2, 2020.

8. Officer, Charles B., and John H. Ryther, "Secondary Sewage Treatment Versus Ocean Outfalls: An Assessment," *Science* 197 (1977): 1056–60.

9. Fennel, Katja, and Jeremy M. Testa, "Biogeochemical Controls on Coastal Hypoxia," *Annual Review of Marine Science* 11 (2019): 105–30.

10. Buchheister, A., C. F. Bonzek, J. Gartland, and R. J. Latour, "Patterns and Drivers of the Demersal Fish Community of Chesapeake Bay," *Marine Ecology Progress Series* 481 (2013): 161–80.

11. Seliger, H. H., J. A. Boggs, and W. H. Biggley, "Catastrophic Anoxia in the Chesapeake Bay in 1984," *Science* 228 (1985): 70–73.

12. Stickle, William B., Martin A. Kapper, Li-Lian Liu, Erich Gnaiger, and Shiao Y. Wang, "Metabolic Adaptations of Several Species of Crustaceans and Molluscs to Hypoxia: Tolerance and Microcalorimetric Studies," *Biological Bulletin* 177 (1989): 303–12.

13. Breitburg, Denise, Lisa A. Levin, Andreas Oschlies, Marilaure Grégoire, Francisco P. Chavez, Daniel J. Conley, Véronique Garçon, et al., "Declining Oxygen in the Global Ocean and Coastal Waters," *Science* 359 (2018): https://doi.org/10.1126/science.aam7240.

14. Thomas, Peter, Md. Saydur Rahman, Matthew E. Picha, and Wenxian Tan, "Impaired Gamete Production and Viability in Atlantic Croaker Collected Throughout the 20,000 km^2 Hypoxic Region in the Northern Gulf of Mexico," *Marine Pollution Bulletin* 101 (2015): 182–92.

15. Saaresranta, T., and O. Polo, "Sleep-Disordered Breathing and Hormones," *European Respiratory Journal* 22 (2003): 161–72.

16. Keppel, A. G., D. L. Breitburg, G. H. Wikfors, R. B. Burrell, and V. M. Clark, "Effects of Co-Varying Diel-Cycling Hypoxia and pH on Disease Susceptibility in the Eastern Oyster *Crassostrea Virginica*," *Marine Ecology Progress Series* 538 (2015): 169–83.

17. Breitburg, Denise L., Darryl W. Hondorp, Lori A. Davias, and Robert J. Diaz, "Hypoxia, Nitrogen, and Fisheries: Integrating Effects across Local and Global Landscapes," *Annual Review of Marine Science* 1 (2009): 329–49.

18. McCormick, Lillian R., and Lisa A. Levin, "Physiological and Ecological Implications of Ocean Deoxygenation for Vision in Marine Organisms," *Philosophical Transactions of the Royal Society A: Mathematical, Physical and Engineering Sciences* 375 (2017): https://doi.org/10.1098/rsta.2016.0322.

19. Wang, Simon Yuan, Karen Lau, Keng-Po Lai, Jiang-Wen Zhang, Anna Chung-Kwan Tse, Jing-Woei Li, Yin Tong et al., "Hypoxia Causes Transgenerational Impairments in Reproduction of Fish," *Nature Communications* 7 (2016): 12114.

20. Fonselius, Stig, and Jorge Valderrama, "One Hundred Years of Hydrographic Measurements in the Baltic Sea," *Journal of Sea Research* 49 (2003): 229–41.

21. Pettersson, Otto, "A Review of the Swedish Hydrographic Research in the Baltic and North Sea," *Scottish Geographical Magazine* 10 (1894): 282.

22. Fonselius, Stig H., *Hydrography of the Baltic Deep Basins III*, Fishery Board of Sweden, Series Hydrography, Report No. 23 (1969), p. 50.

23. Fonselius, *Hydrography of the Baltic*.

24. Reusch, Thorsten B. H., Jan Dierking, Helen C. Andersson, Erik Bonsdorff, Jacob Carstensen, Michele Casini, Mikolaj Czajkowski, et al., "The Baltic Sea as a Time Machine for the Future Coastal Ocean," *Science Advances* 4 (2018): https://doi.org/10.1126/sciadv.aar8195.

25. Carstensen, Jacob, and Daniel J. Conley, "Baltic Sea Hypoxia Takes Many Shapes and Sizes," *Limnology and Oceanography Bulletin* 28 (2019): 125–29.

26. Kahru, M., and R. Elmgren, "Multidecadal Time Series of Satellite-Detected Accumulations of Cyanobacteria in the Baltic Sea," *Biogeosciences* 11 (2014): 3619–33.

27. Karjalainen, Miina, Jonna Engström-Öst, Samuli Korpinen, Heikki Peltonen, Jari-Pekka Pääkkönen, Sanna Rönkkönen, Sanna Suikkanen, et al., "Ecosystem Consequences of Cyanobacteria in the Northern Baltic Sea," *Ambio* 36 (2007): 195–202.

28. Wilford, John Noble, "Geologists Link Black Sea Deluge to Farming's Rise," *New York Times*, December 17, 1996, pp. 1, sec. C.

29. Yanchilina, Anastasia G., William B. F. Ryan, Jerry F. McManus, Petko Dimitrov, Dimitar Dimitrov, Krasimira Slavova, and Mariana Filipova-Marinova, "Compilation of Geophysical, Geochronological, and Geochemical Evidence Indicates a Rapid Mediterranean-Derived Submergence of the Black Sea's Shelf and Subsequent Substantial Salinification in the Early Holocene," *Marine Geology* 383 (2017): 14–34.

30. Bailey, Douglass W., "Holocene Changes in the Level of the Black Sea: Consequences at a Human Scale," in *The Black Sea Flood Question: Changes in Coastline, Climate, and Human Settlement*, edited by Valentina Yanko-Hombach, et al. (Dordrecht: Springer Netherlands, 2007), pp. 515–36.

31. Schuiling, Roelof Dirk, Richard B. Cathcart, Viorel Badescu, Dragos Isvoranu, and Efim Pelinovsky, "Asteroid Impact in the Black Sea. Death by Drowning or Asphyxiation?," *Natural Hazards* 40 (2007): 327–38.

32. Tolmazin, D., "Changing Coastal Oceanography of the Black Sea. I: Northwestern Shelf," *Progress in Oceanography* 15 (1985): 217–76.

33. Rabalais, N. N., R. E. Turner, and W. J. Wiseman, "Gulf of Mexico Hypoxia, aka 'the Dead Zone,'" *Annual Review of Ecology and Systematics* 33 (2002): 235–63.

34. Chen, Chung-Chi, Gwo-Ching Gong, and Fuh-Kwo Shiah, "Hypoxia in the East China Sea: One of the Largest Coastal Low-Oxygen Areas in the World," *Marine Environmental Research* 64 (2007): 399–408.

35. Altieri, Andrew H., Seamus B. Harrison, Janina Seemann, Rachel Collin, Robert J. Diaz, and Nancy Knowlton, "Tropical Dead Zones and Mass Mortalities on Coral Reefs," *Proceedings of the National Academy of Sciences* 114 (2017): 3660–65.

Chapter 3

1. https://boards.cruisecritic.com/showthread.php?t=1792961, accessed June 28, 2018.

2. Shaffer, Gary P., John W. Day Jr, Sarah Mack, G. Paul Kemp, Ivor van Heerden, Michael A. Poirrier, Karen A. Westphal, et al., "The MRGO Navigation Project: A

Massive Human-Induced Environmental, Economic, and Storm Disaster," *Journal of Coastal Research* 54 (2009): 206–24.

3. Turner, R. E., and R. L. Allen, "Bottom Water Oxygen Concentration in the Mississippi River Delta Bight," *Contributions in Marine Science* 25 (1982): 161–72.

4. Grasshoff, K., "The Hydrochemistry of Landlocked Basins and Fjords," in *Chemical Oceanography*, edited by J. P. Riley and G. Skirrow (London: Academic Press, 1975), pp. 455–597.

5. Cutler, David, and Grant Miller, "The Role of Public Health Improvements in Health Advances: The Twentieth-Century United States," *Demography* 42 (2005): 1–22.

6. Boesch, D. F., "Implications of Oxygen Depletion on the Continental Shelf of the Northern Gulf of Mexico," *Coastal Ocean Pollution Assessment News* 2 (1983): 25–28.

7. https://www.vims.edu/features/people/boesch_d.php, accessed October 14, 2019.

8. https://lumcon.edu/history/, accessed June 8, 2018.

9. Boesch, Don. Interview with author, Cambridge, MD, August 24, 2016.

10. Rayl, A. J. S., "Profile of Nancy N. Rabalais, Ph.D.," https://gustavus.edu/events/nobelconference/2009/rabalais-profile.php, accessed March 17, 2020.

11. Ferber, Dan, "The Dead Zone's Fiercest Crusaders," *Science* 291 (2001): 970–71.

12. Rabalais, N. N., R. E. Turner, and W. J. Wiseman, "Gulf of Mexico Hypoxia, Aka 'the Dead Zone,'" *Annual Review of Ecology and Systematics* 33 (2002): 235–63.

13. Laurent, Arnaud, and Katja Fennel, "Time-Evolving, Spatially Explicit Forecasts of the Northern Gulf of Mexico Hypoxic Zone," *Environmental Science & Technology* 53 (2019): 14449–58.

14. Stewart-Abernathy, Leslie C., and Hester A. Davis, eds., *Ghost Boats on the Mississippi: Discovering Our Working Past* (Fayetteville: Arkansas Archeological Survey, 2002).

15. https://www.nwrfc.noaa.gov/floods/papers/oh_2/great.htm, accessed October 2, 2019.

16. http://www.mvn.usace.army.mil/Missions/Mississippi-River-Flood-Control/Bonnet-Carre-Spillway-Overview/Spillway-Operation-Information/, accessed June 14, 2018.

17. Matli, V. Rohith Reddy, Shiqi Fang, Joseph Guinness, Nancy N. Rabalais, J. Kevin Craig, and Daniel R. Obenour, "Space-Time Geostatistical Assessment of Hypoxia in the Northern Gulf of Mexico," *Environmental Science & Technology* 52 (2018): 12484–93.

18. There are actually even fewer samples than 35. No data are available for 1989 and 2016, according to https://gulfhypoxia.net/research/shelfwide-cruises/, accessed March 12, 2020.

19. Nelsen, Terry A., Pat Blackwelder, Terri Hood, Brent McKee, Nidia Romer, Carlos Alvarez-Zarikian, and Simone Metz, "Time-Based Correlation of Biogenic, Lithogenic and Authigenic Sediment Components with Anthropogenic Inputs in the Gulf of Mexico NECOP Study Area," *Estuaries* 17 (1994): 873–85.

20. Pearson, Paul N., "Oxygen Isotopes in Foraminifera: Overview and Historical Review," *The Paleontological Society Papers* 18 (2012): 1–38.

21. Vénec-Peyré, M.-T., and A. Bartolini, "Alcide D'Orbigny and the Paris Foraminiferal Collection," in *Landmarks in Foraminiferal Micropalaeontology: History and*

Development, edited by A. J. Bowden, F. J. Gregory, and Andrew S. Henderson (London: The Geological Society, 2013), pp. 9–22.

22. Miller, C. Giles, "A Brief History of Modeling Foraminifera: From D'Orbigny to Zheng Shouyi," in *Landmarks in Foraminiferal Micropalaeontology: History and Development*, edited by A. J. Bowden, F. J. Gregory, and Andrew S. Henderson (London: The Geological Society, 2013), pp. 337–49.

23. Galloway, W. E., D. G. Bebout, W. L. Fisher Jr., J. G. Dunlap, R. C. Cabrera-Castro, J. E. Lugo-Rivera, and T. M. Scott, "Cenozoic," in *The Gulf of Mexico Basin*, edited by Amos Salvador, Geology of North America (Boulder, CO: Geological Society of America, 1991), pp. 245–324.

24. Sen Gupta, Barun K., R. Eugene Turner, and Nancy N. Rabalais, "Seasonal Oxygen Depletion in Continental-Shelf Waters of Louisiana: Historical Record of Benthic Foraminifers," *Geology* 24 (1996): 227–30.

25. Swarzenski, P. W., P. L. Campbell, L. E. Osterman, and R. Z. Poore, "A 1000-Year Sediment Record of Recurring Hypoxia Off the Mississippi River: The Potential Role of Terrestrially-Derived Organic Matter Inputs," *Marine Chemistry* 109 (2008): 130–42.

26. Osterman, L. E., R. Z. Poore, P. W. Swarzenski, and R. E. Turner, "Reconstructing a 180 Yr Record of Natural and Anthropogenic Induced Low-Oxygen Conditions from Louisiana Continental Shelf Sediments," *Geology* 33 (2005): 329–32.

27. Osterman, L. E., R. Z. Poore, and P. W. Swarzenski, "The Last 1000 Years of Natural and Anthropogenic Low-Oxygen Bottom-Water on the Louisiana Shelf, Gulf of Mexico," *Marine Micropaleontology* 66 (2008): 291–303.

28. Djakovac, Tamara, Nastjenjka Supić, Fabrizio Bernardi Aubry, Danilo Degobbis, and Michele Giani, "Mechanisms of Hypoxia Frequency Changes in the Northern Adriatic Sea During the Period 1972–2012," *Journal of Marine Systems* 141 (2015): 179–89.

29. Justić, Dubravko, Tarzan Legović, and Laura Rottini-Sandrini, "Trends in Oxygen Content 1911–1984 and Occurrence of Benthic Mortality in the Northern Adriatic Sea," *Estuarine, Coastal and Shelf Science* 25 (1987): 435–45.

30. Turner, Gene. Interview with author, Baton Rouge, LA, January 31, 2019.

31. Stachowitsch, Michael, Neda Fanuko, and Marjan Richter, "Mucus Aggregates in the Adriatic Sea: An Overview of Stages and Occurrences," *Marine Ecology* 11 (1990): 327–50.

32. Justić, Dubravko. Interview with author, Baton Rouge, LA, January 30, 2019.

33. Kralj, M., M. Lipizer, B. Čermelj, M. Celio, C. Fabbro, F. Brunetti, J. Francé, et al., "Hypoxia and Dissolved Oxygen Trends in the Northeastern Adriatic Sea (Gulf of Trieste)," *Deep Sea Research Part II: Topical Studies in Oceanography* 164 (2019): 74–88.

34. Tomašových, Adam, Ivo Gallmetzer, Alexandra Haselmair, Darrell S. Kaufman, Jelena Vidović, and Martiin Zuschin, "Stratigraphic Unmixing Reveals Repeated Hypoxia Events over the Past 500 Yr in the Northern Adriatic Sea," *Geology* 45 (2017): 363–66.

35. Barmawidjaja, D. M., G. J. van der Zwaan, F. J. Jorissen, and S. Puskaric, "150 Years of Eutrophication in the Northern Adriatic Sea: Evidence from a Benthic Foraminiferal Record," *Marine Geology* 122 (1995): 367–84.

36. Fonselius, Stig H., *Hydrography of the Baltic Deep Basins III*, Fishery Board of Sweden, Series Hydrography, Report No. 23 (1969), p. 50.

37. Elmgren, Ragnar, "Man's Impact on the Ecosystem of the Baltic Sea: Energy Flows Today and at the Turn of the Century," *Ambio* 18 (1989): 326–32.

38. Conley, Daniel J., Jacob Carstensen, Juris Aigars, Philip Axe, Erik Bonsdorff, Tatjana Eremina, Britt-Marie Haahti, et al., "Hypoxia Is Increasing in the Coastal Zone of the Baltic Sea," *Environmental Science & Technology* 45 (2011): 6777–83.

39. Carstensen, Jacob, Jesper H. Andersen, Bo G. Gustafsson, and Daniel J. Conley, "Deoxygenation of the Baltic Sea During the Last Century," *Proceedings of the National Academy of Sciences* 111 (2014): 5628–33.

40. Diaz, Robert J., Rutger Rosenberg, and Kersey Sturdivant, "Hypoxia in Estuaries and Semi-Enclosed Seas," in *Ocean Deoxygenation: Everyone's Problem: Causes, Impacts, Consequences and Solutions*, edited by D. Laffoley and J. M. Baxter (Gland, Switzerland: IUCN, 2019), pp. 85–102.

Chapter 4

1. Smith, Lawson M., and Brien R. Winkley, "The Response of the Lower Mississippi River to River Engineering," *Engineering Geology* 45 (1996): 433–55.

2. Munoz, Samuel E., Liviu Giosan, Matthew D. Therrell, Jonathan W. F. Remo, Zhixiong Shen, Richard M. Sullivan, Charlotte Wiman, et al., "Climatic Control of Mississippi River Flood Hazard Amplified by River Engineering," *Nature* 556 (2018): 95.

3. Turner, R. Eugene, and Nancy N. Rabalais, "Changes in Mississippi River Water Quality This Century: Implications for Coastal Food Webs," *BioScience* 41 (1991): 140–47.

4. Twain, Mark, *Life on the Mississippi* (Boston: James R. Osgood and Company, 1883), p. 208.

5. Easterling, D. R., K. E. Kunkel, J. R. Arnold, T. Knutson, A. N. LeGrande, L. R. Leung, R. S. Vose, et al., "Precipitation Change in the United States," in *Climate Science Special Report: Fourth National Climate Assessment*, edited by D. J. Wuebbles et al. (Washington, DC: US Global Change Research Program, 2017), pp. 207–30.

6. Liu, Mingliang, Hanqin Tian, Qichun Yang, Jia Yang, Xia Song, Steven E. Lohrenz, and Wei-Jun Cai, "Long-Term Trends in Evapotranspiration and Runoff over the Drainage Basins of the Gulf of Mexico During 1901–2008," *Water Resources Research* 49 (2013): 1988–2012.

7. Zhang, Y. K., and K. E. Schilling, "Increasing Streamflow and Baseflow in Mississippi River since the 1940s: Effect of Land Use Change," *Journal of Hydrology* 324 (2006): 412–22.

8. Tao, Bo, Hanqin Tian, Wei Ren, Jia Yang, Qichun Yang, Ruoying He, Weijun Cai, et al., "Increasing Mississippi River Discharge Throughout the 21st Century Influenced

by Changes in Climate, Land Use, and Atmospheric CO$_2$," *Geophysical Research Letters* 41 (2014): 4978–86.

9. Barnett, James F., *Beyond Control: The Mississippi River's New Channel to the Gulf of Mexico*, ed. Carl A. Brasseaux and Donald W. Davis, America's Third Coast (Jackson: University Press of Mississippi, 2017).

10. McPhee, John, *The Control of Nature* (New York: Farrar, Straus, Giroux, 1989), p. 8.

11. Bianchi, T. S., S. F. DiMarco, J. H. Cowan Jr, R. D. Hetland, P. Chapman, J. W. Day, and M. A. Allison, "The Science of Hypoxia in the Northern Gulf of Mexico: A Review," *Science of The Total Environment* 408 (2010): 1471–84.

12. Turner, R. Eugene, and Nancy N. Rabalais, "Coastal Eutrophication near the Mississippi River Delta," *Nature* 368 (1994): 619–21.

13. Turner, R. Eugene, and Nancy N. Rabalais, "Linking Landscape and Water Quality in the Mississippi River Basin for 200 Years," *BioScience* 53 (2003): 563–72.

14. Whitney, Gordon Graham, *From Coastal Wilderness to Fruited Plain: A History of Environmental Change in Temperate North America, 1500 to the Present* (Cambridge: Cambridge University Press, 1994), p. 227.

15. Rabalais, Nancy N., and R. Eugene Turner, "Gulf of Mexico Hypoxia: Past, Present, and Future," *Limnology and Oceanography Bulletin* 28 (2019): 117–24.

16. Parsons, Michael L., and Quay Dortch, "Sedimentological Evidence of an Increase in *Pseudo-Nitzschia* (Bacillariophyceae) Abundance in Response to Coastal Eutrophication," *Limnology and Oceanography* 47 (2002): 551–58.

17. Feng, Y., S. F. DiMarco, and G. A. Jackson, "Relative Role of Wind Forcing and Riverine Nutrient Input on the Extent of Hypoxia in the Northern Gulf of Mexico," *Geophysical Research Letters* 39 (2012): https://doi.org/10.1029/2012gl051192.

18. https://gulfhypoxia.net/research/shelfwide-cruise, accessed September 19, 2019.

19. Rabalais, N. N., R. E. Turner, B. K. Sen Gupta, D. F. Boesch, P. Chapman, and M. C. Murrell, "Hypoxia in the Northern Gulf of Mexico: Does the Science Support the Plan to Reduce, Mitigate, and Control Hypoxia?," *Estuaries and Coasts* 30 (2007): 753–72.

20. Bianchi, Thomas. Telephone interview with author, November 21, 2019.

21. Bianchi, Thomas S., Laura A. Wysocki, Kathryn M. Schreiner, Timothy R. Filley, D. Reide Corbett, and Alexander S. Kolker, "Sources of Terrestrial Organic Carbon in the Mississippi Plume Region: Evidence for the Importance of Coastal Marsh Inputs," *Aquatic Geochemistry* 17 (2011): 431–56.

22. Couvillion, Brady R., Holly Beck, Donald Schoolmaster, and Michelle Fischer, "Land Area Change in Coastal Louisiana (1932 to 2016)," (2017), http://pubs.er.usgs.gov/publication/sim3381.

23. Wang, Hongjie, Xinping Hu, Nancy N. Rabalais, and Jay Brandes, "Drivers of Oxygen Consumption in the Northern Gulf of Mexico Hypoxic Waters—A Stable Carbon Isotope Perspective," *Geophysical Research Letters* 45 (2018): 10,528–10,538.

24. Obenour, Daniel R., Anna M. Michalak, Yuntao Zhou, and Donald Scavia, "Quantifying the Impacts of Stratification and Nutrient Loading on Hypoxia in the Northern Gulf of Mexico," *Environmental Science & Technology* 46 (2012): 5489–96.

25. Bianchi, Thomas S., "Controlling Hypoxia on the U.S. Louisiana Shelf: Beyond the Nutrient-Centric View," *Eos* 89 (2008): 236–37.

26. Bianchi, Thomas S., "New Approaches to the Gulf Hypoxia Problem," *Eos* 91 (2010): 173–74.

27. Boesch, Donald. Conversation with author, Mobile, AL, November 7, 2019; and email to author, November 25, 2019.

28. Harding, L. W., C. L. Gallegos, E. S. Perry, W. D. Miller, J. E. Adolf, M. E. Mallonee, and H. W. Paerl, "Long-Term Trends of Nutrients and Phytoplankton in Chesapeake Bay," *Estuaries and Coasts* 39 (2016): 664–81.

29. Rice, Karen C., Douglas L. Moyer, and Aaron L. Mills, "Riverine Discharges to Chesapeake Bay: Analysis of Long-Term (1927–2014) Records and Implications for Future Flows in the Chesapeake Bay Basin," *Journal of Environmental Management* 204 (2017): 246–54.

30. Najjar, Raymond G., Christopher R. Pyke, Mary Beth Adams, Denise Breitburg, Carl Hershner, Michael Kemp, Robert Howarth, et al., "Potential Climate-Change Impacts on the Chesapeake Bay," *Estuarine, Coastal and Shelf Science* 86 (2010): 1–20.

31. Harding et al., "Long-Term Trends."

32. Brush, Grace S., "Historical Land Use, Nitrogen, and Coastal Eutrophication: A Paleoecological Perspective," *Estuaries and Coasts* 32 (2009): 18–28.

33. Pellicer, Jaume, Oriane Hidalgo, Sònia Garcia, Teresa Garnatje, Aleksandr A. Korobkov, Joan Vallès, and Joan Martín, "Palynological Study of *Ajania* and Related Genera (Asteraceae, Anthemideae)," *Botanical Journal of the Linnean Society* 161 (2009): 171–89.

34. Brush, Grace S., and William B. Hilgartner, "Paleoecology of Submerged Macrophytes in the Upper Chesapeake Bay," *Ecological Monographs* 70 (2000): 645–67.

35. Barmawidjaja, D. M., G. J. van der Zwaan, F. J. Jorissen, and S. Puskaric, "150 Years of Eutrophication in the Northern Adriatic Sea: Evidence from a Benthic Foraminiferal Record," *Marine Geology* 122 (1995): 367–84.

36. Capet, A., E. V. Stanev, J. M. Beckers, J. W. Murray, and M. Grégoire, "Decline of the Black Sea Oxygen Inventory," *Biogeosciences* 13 (2016): 1287–97.

37. Li, X. A., Z. M. Yu, X. X. Song, X. H. Cao, and Y. Q. Yuan, "The Seasonal Characteristics of Dissolved Oxygen Distribution and Hypoxia in the Changjiang Estuary," *Journal of Coastal Research* 27 (2011): 52–62.

38. Fonselius, Stig H., "Hydrography of the Baltic Deep Basins III," *Fishery Board of Sweden, Series Hydrography,* Report Number 23 (1969): 1–97.

39. Larsson, Ulf, Ragnar Elmgren, and Fredrik Wulff, "Eutrophication and the Baltic Sea: Causes and Consequences," *Ambio* 14 (1985): 9–14.

40. Email from Conley to author, September 1, 2016; and interview with author, Lund, Sweden, May 23, 2019.

41. Conley, Daniel J., Christoph Humborg, Lars Rahm, Oleg P. Savchuk, and Fredrik Wulff, "Hypoxia in the Baltic Sea and Basin-Scale Changes in Phosphorus Biogeochemistry," *Environmental Science & Technology* 36 (2002): 5315–20.

42. Mohrholz, V., M. Naumann, G. Nausch, S. Krüger, and U. Gräwe, "Fresh Oxygen for the Baltic Sea—An Exceptional Saline Inflow after a Decade of Stagnation," *Journal of Marine Systems* 148 (2015): 152–66.

43. Meier, H. E. Markus, Germo Väli, Michael Naumann, Kari Eilola, and Claudia Frauen, "Recently Accelerated Oxygen Consumption Rates Amplify Deoxygenation in the Baltic Sea," *Journal of Geophysical Research: Oceans* 123 (2018): 3227–40.

44. Mohrholz, Volker, "Major Baltic Inflow Statistics—Revised," *Frontiers in Marine Science* 5 (2018): https://doi.org/10.3389/fmars.2018.00384.

45. Zillén, Lovisa, Daniel J. Conley, Thomas Andrén, Elinor Andrén, and Svante Björck, "Past Occurrences of Hypoxia in the Baltic Sea and the Role of Climate Variability, Environmental Change and Human Impact," *Earth-Science Reviews* 91 (2008): 77–92.

46. Zillén et al., "Past Occurrences of Hypoxia."

47. Norbäck Ivarsson, Lena, Thomas Andrén, Matthias Moros, Thorbjørn Joest Andersen, Mikael Lönn, and Elinor Andrén, "Baltic Sea Coastal Eutrophication in a Thousand Year Perspective," *Frontiers in Environmental Science* 7 (2019): https://doi.org/10.3389/fenvs.2019.00088.

48. Ruddiman, W. F., *Earth's Climate: Past and Future* (New York: W.H. Freeman, 2001).

Chapter 5

1. These calculations assume $C:O_2$ equal to 1 and that the algae make organic material with a C:N:P ratio of 106:16:1, that is, the Redfield ratio. The range given in the main text depends on whether nitrogen or phosphorus is limiting.

2. Robertson, Dale M., and David A. Saad, "Sparrow Models Used to Understand Nutrient Sources in the Mississippi/Atchafalaya River Basin," *Journal of Environmental Quality* 42 (2013): 1422–40.

3. HELCOM, "Sources and Pathways of Nutrients to the Baltic Sea," (Helsinki Commission: Helsinki, Finland, 2018) https://helcom.fi/wp-content/uploads/2019/08/Overview-of-nutrient-recycling-in-the-Baltic-Sea-countries.pdf.

4. Popovici, Mihaela, "Nutrient Management in the Danube River Basin," in *The Danube River Basin*, edited by Igor Liska (Berlin: Springer Berlin Heidelberg, 2015), pp. 23–38.

5. https://www.epa.gov/chesapeake-bay-tmdl, accessed July 20, 2018.

6. USDA Foreign Agricultural Service data for 2020, from https://www.fas.usda.gov, accessed September 22, 2020.

7. Map m13190 from the US Department of Agriculture, National Resources Conservation Service.

8. The city claimed the soybean title mainly to honor one of its prominent companies, A. E. Staley Manufacturing, which was an early promotor of soybeans, http://www.soyinfocenter.com/HSS/ae_staley_manufacturing.php, accessed July 25, 2018.

9. Data from USDA National Agricultural Statistics Service (https://www.nass.usda.gov/), accessed August 9, 2018.

10. Busacca, Alan J., James E. Begét, Helaine W. Markewich, Daniel R. Muhs, Nicholas Lancaster, and Mark R. Sweeney, "Eolian Sediments," in *The Quaternary Period in the United States*, edited by Alan R. Gillespie, Stephen C. Porter, and Brian F. Atwater,

Developments in Quaternary Science (Amsterdam, Netherlands: Elsevier, 2004), pp. 275–309.

11. Leopold, Aldo, "Lakes in Relation to Terrestrial Life Patterns," in *A Symposium on Hydrobiology*, edited by James G. Needham (Madison: The University of Wisconsin Press, 1941), p. 18.

12. Yu, Zhen, and Chaoqun Lu, "Historical Cropland Expansion and Abandonment in the Continental U.S. During 1850 to 2016," *Global Ecology and Biogeography* 27 (2018): 322–33.

13. Data from USDA Census of Agriculture. The 1860 and 1950 data are available at http://agcensus.mannlib.cornell.edu/AgCensus/.

14. Yu and Lu, "Historical Cropland Expansion."

15. McIsaac, Gregory, and Mark B. David, "On the Need for Consistent and Comprehensive Treatment of the N Cycle," *Science of The Total Environment* 305 (2003): 249–55.

16. Egli, D. B., "Comparison of Corn and Soybean Yields in the United States: Historical Trends and Future Prospects," *Agronomy Journal* 100 (2008): S-79–S-88.

17. Veseth, Johannes Olsen, and Gunnar J. Malmin, "I Would I Were on the Mississippi: An Emigrant Dialogue," in *Norwegian Emigrant Songs and Ballads*, edited by Theodore C. Blegen and Martin B. Ruud (Minneapolis: University of Minnesota Press, 1936), pp. 67–74.

18. Anderson, J. L., *Industrializing the Corn Belt: Agriculture, Technology, and Environment, 1945-1972* (DeKalb: Northern Illinois University Press, 2009), p. 52.

19. Beers, Z. H., "Development of the Fertilizer Industry in the Middle Western States," *Fertilizer Review* 26 (1951): 4–7.

20. Anonymous, "Dollar Corn Makes Fertilizer Pay," *Wallaces Farmer*, January 15, 1944, pp. 46–47.

21. Data from USDA. The 1950 data are from Statistical Bulletin No. 191, while the 1979 data are from "1980 Fertilizer Situation," FS10.

22. Cao, P., C. Lu, and Z. Yu, "Historical Nitrogen Fertilizer Use in Agricultural Ecosystems of the Contiguous United States During 1850–2015: Application Rate, Timing, and Fertilizer Types," *Earth System Science Data* 10 (2018): 969–84.

23. https://www.epa.gov/nutrient-policy-data/commercial-fertilizer-purchased, accessed August 31, 2018.

24. Stewart, W. M., D. W. Dibb, A. E. Johnston, and T. J. Smyth, "The Contribution of Commercial Fertilizer Nutrients to Food Production," *Agronomy Journal* 97 (2005): 1–6.

25. Lu, Chaoqun, Jien Zhang, Peiyu Cao, and Jerry L. Hatfield, "Are We Getting Better in Using Nitrogen?: Variations in Nitrogen Use Efficiency of Two Cereal Crops across the United States," *Earth's Future* 7 (2019): 939–52.

26. Dhillon, Jagmandeep, Guilherme Torres, Ethan Driver, Bruno Figueiredo, and William R. Raun, "World Phosphorus Use Efficiency in Cereal Crops," *Agronomy Journal* 109 (2017): 1670–77.

27. Smil, Vaclav, *Enriching the Earth: Fritz Haber, Carl Bosch, and the Transformation of World Food Production* (Cambridge, MA: MIT Press, 2001).

28. Smil, Vaclav, "Phosphorus in the Environment: Natural Flows and Human Interferences," *Annual Review of Energy and the Environment* 25 (2000): 53–88.

29. Christianson, L. E., R. D. Harmel, D. Smith, M. R. Williams, and K. King, "Assessment and Synthesis of 50 Years of Published Drainage Phosphorus Losses," *Journal of Environmental Quality* 45 (2016): 1467–77.

30. Turner, R. Eugene, and Nancy N. Rabalais, "Changes in Mississippi River Water Quality This Century: Implications for Coastal Food Webs," *BioScience* 41 (1991): 140–47.

31. De Cicco, L. A., L. A. Sprague, J. C. Murphy, M. L. Riskin, J. A. Falcone, E. G. Stets, G. P. Oelsner, et al., "Water-Quality and Streamflow Datasets Used in the Weighted Regressions on Time, Discharge, and Season (WRTDS) Models to Determine Trends in the Nation's Rivers and Streams, 1972-2012 ([Ver. 1.1 July 7, 2017]," US Geological Survey 2017).

32. Cao et al., "Historical Nitrogen Fertilizer Use."

33. Howarth, R. W., G. Billen, D. Swaney, A. Townsend, N. Jaworski, K. Lajtha, J. A. Downing, et al., "Regional Nitrogen Budgets and Riverine N & P Fluxes for the Drainages to the North Atlantic Ocean: Natural and Human Influences," *Biogeochemistry* 35 (1996): 75–139.

34. Svanbäck, Annika, Michelle L. McCrackin, Dennis P. Swaney, Helena Linefur, Bo G. Gustafsson, Robert W. Howarth, and Christoph Humborg, "Reducing Agricultural Nutrient Surpluses in a Large Catchment—Links to Livestock Density," *Science of The Total Environment* 648 (2019): 1549–59.

35. Fanelli, Rosemary M., Joel D. Blomquist, and Robert M. Hirsch, "Point Sources and Agricultural Practices Control Spatial-Temporal Patterns of Orthophosphate in Tributaries to Chesapeake Bay," *Science of The Total Environment* 652 (2019): 422–33.

36. Jones, Christopher S., Chad W. Drake, Claire E. Hruby, Keith E. Schilling, and Calvin F. Wolter, "Livestock Manure Driving Stream Nitrate," *Ambio* 48 (2019): 1143–53.

37. Jones, Chris, "Iowa's Real Population," March 14, 2019, https://www.iihr.uiowa.edu/cjones/iowas-real-population, accessed March 17, 2020.

38. Jones et al., "Livestock Manure," p. 1143.

39. Larsson, Ulf, Ragnar Elmgren, and Fredrik Wulff, "Eutrophication and the Baltic Sea: Causes and Consequences," *Ambio* 14 (1985): 9–14.

40. Ferber, Dan, "Keeping the Stygian Waters at Bay," *Science* 291 (2001): 968–73.

41. Rabalais, Nancy N., R. E. Turner, Dubravko Justić, Quay Dortch, and William J. Wiseman Jr., "Characterization of Hypoxia: Topic I Report for the Integrated Assessment on Hypoxia in the Gulf of Mexico," NOAA (Silver Spring, MD, 1999).

42. Goolsby, Donald A., William A. Battaglin, Gregory B. Lawrence, Richard S. Artz, Brent T. Aulenbach, Richard P. Hooper, Dennis R. Keeney, et al., "Flux and Sources of Nutrients in the Mississippi–Atchafalaya River Basin: Topic 3 Report for the Integrated Assessment on Hypoxia in the Gulf of Mexico," NOAA (Silver Spring, MD, 1999).

43. Ferber, Dan, "The Dead Zone's Fiercest Crusaders," *Science* 291 (2001): 970–71.

44. Krug, Edward C., and Derek Winstanley, "A Contribution to the Characterization of Illinois Reference/Background Conditions for Setting Nitrogen Criteria for Surface Waters in Illinois," Illinois Department of Natural Resources (Champaign, IL, 2000), p. 10.

45. Ferber, "The Dead Zone's Fiercest Crusaders."

46. Scavia, Don. Interview with author, Newark, DE, March 14, 2019.

47. Bailenson, Stephanie, email to author, February 10, 2020.

48. https://www.congress.gov/bill/105th-congress/senate-bill/1480, accessed September 20, 2019.

49. https://www.chesapeakeprogress.com/clean-water/watershed-implementation-plans, accessed March 21, 2019.

50. HELCOM, "Sources and Pathways of Nutrients."

51. Mee, Laurence, D., "Reviving Dead Zones," *Scientific American* 295 (2006): 78–85.

52. Capet, A., J. M. Beckers, and M. Grégoire, "Drivers, Mechanisms and Long-Term Variability of Seasonal Hypoxia on the Black Sea Northwestern Shelf—Is There Any Recovery after Eutrophication?," *Biogeosciences* 10 (2013): 3943–62.

53. He, Qiang, Mark D. Bertness, John F. Bruno, Bo Li, Guoqian Chen, Tyler C. Coverdale, Andrew H. Altieri, et al., "Economic Development and Coastal Ecosystem Change in China," *Scientific Reports* 4 (2014): 5995.

54. Liu, Yunhua, and Xiaobing Wang, "Technological Progress and Chinese Agricultural Growth in the 1990s," *China Economic Review* 16 (2005): 419–40.

55. Lu, C., and H. Tian, "Global Nitrogen and Phosphorus Fertilizer Use for Agriculture Production in the Past Half Century: Shifted Hot Spots and Nutrient Imbalance," *Earth System Science Data* 9 (2017): 181–92.

56. Zhang, Xin, Eric A. Davidson, Denise L. Mauzerall, Timothy D. Searchinger, Patrice Dumas, and Ye Shen, "Managing Nitrogen for Sustainable Development," *Nature* 528 (2015): 51–59.

57. Chen, Xi, Maryna Strokal, Michelle T. H. Van Vliet, John Stuiver, Mengru Wang, Zhaohai Bai, Lin Ma, et al., "Multi-Scale Modeling of Nutrient Pollution in the Rivers of China," *Environmental Science & Technology* 53 (2019): 9614–25.

58. Liu, Dongyan, John K. Keesing, Peimin He, Zongling Wang, Yajun Shi, and Yujue Wang, "The World's Largest Macroalgal Bloom in the Yellow Sea, China: Formation and Implications," *Estuarine, Coastal and Shelf Science* 129 (2013): 2–10.

59. Qian, Wei, Jianping Gan, Jinwen Liu, Biyan He, Zhongming Lu, Xianghui Guo, Deli Wang, et al., "Current Status of Emerging Hypoxia in a Eutrophic Estuary: The Lower Reach of the Pearl River Estuary, China," *Estuarine, Coastal and Shelf Science* 205 (2018): 58–67.

60. Zhu, Zhuo-Yi, Hui Wu, Su-Mei Liu, Ying Wu, Da-Ji Huang, Jing Zhang, and Guo-Sen Zhang, "Hypoxia Off the Changjiang (Yangtze River) Estuary and in the Adjacent East China Sea: Quantitative Approaches to Estimating the Tidal Impact and Nutrient Regeneration," *Marine Pollution Bulletin* 125 (2017): 103–14.

61. Yu, ChaoQing, Xiao Huang, Han Chen, H. Charles J. Godfray, Jonathon S. Wright, Jim W. Hall, Peng Gong, et al., "Managing Nitrogen to Restore Water Quality in China," *Nature* 567 (2019): 516–20.

Chapter 6

1. Brock, W. H., *Justus von Liebig: The Chemical Gatekeeper* (Cambridge: Cambridge University Press, 1997).

2. Davy, Humphry, *Elements of Agricultural Chemistry, in a Course of Lectures for the Board of Agriculture* (New York: Eastburn, Kirk & Co., 1815), p. 270.

3. Liebig, Justus Freiherr von, *Chemistry in Its Applications to Agriculture and Physiology*. 4th ed., rev. and enl. ed. (London: Taylor and Walton, 1847).

4. Davy, Humphry, *Elements of Agricultural Chemistry*.

5. van der Ploeg, R. R., W. Böhm, and M. B. Kirkham, "On the Origin of the Theory of Mineral Nutrition of Plants and the Law of the Minimum," *History of Soil Science* 63 (1999): 1055–62.

6. https://minerals.usgs.gov/minerals/pubs/commodity/potash/, accessed November 14, 2018.

7. Brock, *Justus von Liebig*.

8. Halliday, Stephen, *The Great Stink of London: Sir Joseph Bazalgette and the Cleansing of the Victorian Metropolis* (Stroud: Sutton, 1999), p. 109.

9. https://minerals.usgs.gov/minerals/pubs/commodity/phosphate_rock/, accessed May 14, 2020.

10. Brock, *Justus von Liebig*.

11. Liebig, Justus, *Familiar Letters on Chemistry* (London: Printed for Taylor and Walton, 1844), p. 177.

12. Smil, Vaclav, *Enriching the Earth: Fritz Haber, Carl Bosch, and the Transformation of World Food Production* (Cambridge, MA: MIT Press, 2001).

13. Crookes, William, *The Wheat Problem* (London: John Murray, 1899), p. 3.

14. Smil, *Enriching the Earth*.

15. Smil, *Enriching the Earth*, p. 160.

16. Stoltzenberg, Dietrich, *Fritz Haber: Chemist, Nobel Laureate, German, Jew* (Philadelphia, PA: Chemical Heritage Press, 2004).

17. Stoltzenberg, *Fritz Haber*, p. xxii.

18. Stoltzenberg, *Fritz Haber*, p. 247.

19. Stoltzenberg, *Fritz Haber*, p. 277.

20. Brock, *Justus von Liebig*, p. 213.

21. Broecker, Wallace S., and T.-H. Peng, *Tracers in the Sea* (Palisades, NY: Lamont-Doherty Geological Observatory, Columbia University, 1982), p. 275.

22. Kirchman, D. L., *Processes in Microbial Ecology*, 2nd ed. (Oxford: Oxford University Press, 2018), p. 198.

23. Schindler, David W., "The Dilemma of Controlling Cultural Eutrophication of Lakes," *Proceedings of the Royal Society B: Biological Sciences* 279 (2012): 4322–33.

24. Boyd, P. W., T. Jickells, C. S. Law, S. Blain, E. A. Boyle, K. O. Buesseler, K. H. Coale, et al., "Mesoscale Iron Enrichment Experiments 1993–2005: Synthesis and Future Directions," *Science* 315 (2007): 612–17.

25. Oviatt, C., P. Doering, B. Nowicki, L. Reed, J. J. Cole, and J. Frithsen, "An Ecosystem Level Experiment on Nutrient Limitation in Temperate Coastal Marine Environments" *Marine Ecology Progress Series* 116 (1995): 171–79.

26. Tamminen, Timo, and Tom Andersen, "Seasonal Phytoplankton Nutrient Limitation Patterns as Revealed by Bioassays over Baltic Sea Gradients of Salinity and Eutrophication," *Marine Ecology Progress Series* 340 (2007): 121–38.

27. Turner, R. E., and N. N. Rabalais, "Nitrogen and Phosphorus Phytoplankton Growth Limitation in the Northern Gulf of Mexico," *Aquatic Microbial Ecology* 68 (2013): 159–69.

Chapter 7

1. Williams, Peter J. le B., "An Appreciation of Alfred C. Redfield and His Scientific Work" *Limnology and Oceanography Bulletin* 15 (2006): 53–70.

2. Redfield, Alfred C., "On the Proportions of Organic Derivatives in Sea Water and Their Relation to the Composition of Plankton," in *James Johnstone Memorial Volume* (Liverpool: University Press of Liverpool, 1934), pp. 176–92.

3. Cazzolla Gatti, Roberto, "Coronavirus Outbreak Is a Symptom of Gaia's Sickness," *Ecological Modelling* 426 (2020): 109075.

4. Tyrrell, Toby, *On Gaia: A Critical Investigation of the Relationship between Life and Earth* (Princeton, NJ: Princeton University Press, 2013).

5. Redfield, Alfred C., "The Biological Control of Chemical Factors in the Environment," *American Scientist* 46 (1958): 205–21.

6. Anonymous, "Review of Issues Related to Gulf of Mexico Hypoxia" (Atlanta, Georgia: US EPA Region 4, January 2004).

7. Anonymous, "Evaluation of the Role of Nitrogen and Phosphorous in Causing or Contributing to Hypoxia in the Northern Gulf" (Atlanta, Georgia: US EPA Region 4, August 2004).

8. Ferber, Dan, "Dead Zone Fix Not a Dead Issue," *Science* 305 (2004): 1557.

9. Boesch, Donald F., "The Gulf of Mexico's Dead Zone," *Science* 306 (2004): 977–78.

10. Dodds, Walter K., "Nutrients and the "Dead Zone": The Link between Nutrient Ratios and Dissolved Oxygen in the Northern Gulf of Mexico," *Frontiers in Ecology and the Environment* 4 (2006): 211–17.

11. https://www.epa.gov/ms-htf, accessed November 14, 2019.

12. Fennel, Katja. Interview with author, Newark, DE, April 25, 2019.

13. Fennel, Katja, and Jeremy M. Testa, "Biogeochemical Controls on Coastal Hypoxia," *Annual Review of Marine Science* 11 (2019): 105–30.

14. Fennel, K., and A. Laurent, "N and P as Ultimate and Proximate Limiting Nutrients in the Northern Gulf of Mexico: Implications for Hypoxia Reduction Strategies," *Biogeosciences* 15 (2018): 3121–31.

15. Feng, Yang, Steven F. DiMarco, Karthik Balaguru, and Huijie Xue, "Seasonal and Interannual Variability of Areal Extent of the Gulf of Mexico Hypoxia from a Coupled Physical-Biogeochemical Model: A New Implication for Management Practice," *Journal of Geophysical Research: Biogeosciences* 124 (2019): 1939–60.

16. Crawford, John T., Edward G. Stets, and Lori A. Sprague, "Network Controls on Mean and Variance of Nitrate Loads from the Mississippi River to the Gulf of Mexico," *Journal of Environmental Quality* 48 (2019): 1789–99.

17. Boesch, Donald F., "Barriers and Bridges in Abating Coastal Eutrophication," *Frontiers in Marine Science* 6 (2019): https://doi.org/10.3389/fmars.2019.00123.

18. Ptacnik, Robert, Tom Andersen, and Timo Tamminen, "Performance of the Redfield Ratio and a Family of Nutrient Limitation Indicators as Thresholds for Phytoplankton N vs. P Limitation," *Ecosystems* 13 (2010): 1201–14.

19. Rosenberg, R., R. Elmgren, S. Fleischer, P. Jonsson, G. Persson, and H. Dahlin, "Marine Eutrophication Case-Studies in Sweden," *Ambio* 19 (1990): 102–08.

20. André, Axel, Anna Maria Sundin, Linda Linderholm, Istvan Borbas, and Kristina Svinhufvud, "Wastewater Treatment in Sweden 2016," Swedish Environmental Protection Agency (2016), https://www.naturvardsverket.se/Documents/publikationer6400/978-91-620-8809-5.pdf?pid=22471.

21. Boesch, Donald, Robert Heck, Charles O'Melia, D. W. Schindler, and S. P. Seitzinger, "Eutrophication of Swedish Seas," Swedish Environmental Protection Agency (2006), http://www.naturvardsverket.se/documents/publikationer/620-5509-7.pdf.

22. The quote is from the title of a presentation given by D. W. Schindler in 2015 when he accepted the Redfield Award from the Association for Sciences of Limnology and Oceanography. The title had "OK corral" crossed out and overwritten with "N-P lagoon." The presentation is at https://www.youtube.com/watch?v=i1KUSzBRt7U, accessed February 27, 2020.

23. Schindler, David W., and John R. Vallentyne, *The Algal Bowl: Overfertilization of the World's Freshwaters and Estuaries* (Edmonton: University of Alberta Press, 2008).

24. Rolff, Carl, and Tina Elfwing, "Increasing Nitrogen Limitation in the Bothnian Sea, Potentially Caused by Inflow of Phosphate-Rich Water from the Baltic Proper," *Ambio* 44 (2015): 601–11.

25. Email from Daniel Conley to author, November 24, 2019.

26. Email from Ragnar Elmgren to author, December 10, 2019.

27. Vahtera, E., D. J. Conley, B. G. Gustafsson, H. Kuosa, H. Pitkanen, O. P. Savchuk, T. Tamminen, et al., "Internal Ecosystem Feedbacks Enhance Nitrogen-Fixing Cyanobacteria Blooms and Complicate Management in the Baltic Sea," *Ambio* 36 (2007): 186–94.

28. André et al., "Wastewater Treatment in Sweden."

29. Birge, E. A., "Gases Dissolved in the Waters of Wisconsin Lakes," *Transactions of the American Fisheries Society* 35 (1906): 143–63.

30. Birge, E. A., and Chancey Juday, *The Inland Lakes of Wisconsin: The Dissolved Gases of the Water and Their Biological Significance* (Madison, WI: Wisconsin Geological and Natural History Survey, 1911).

31. Jenny, Jean-Philippe, Pierre Francus, Alexandre Normandeau, François Lapointe, Marie-Elodie Perga, Antti Ojala, Arndt Schimmelmann, et al., "Global Spread of Hypoxia in Freshwater Ecosystems During the Last Three Centuries Is Caused by Rising Local Human Pressure," *Global Change Biology* 22 (2016): 1481–89.

32. Sweeney, Robert A., "'Dead' Sea of North America?—Lake Erie in the 1960s and '70s," *Journal of Great Lakes Research* 19 (1993): 198–99.

33. US Department of Agriculture, Economics, Statistics and Market Information System.

34. Scavia, Donald, J. David Allan, Kristin K. Arend, Steven Bartell, Dmitry Beletsky, Nate S. Bosch, Stephen B. Brandt, et al., "Assessing and Addressing the Re-Eutrophication of Lake Erie: Central Basin Hypoxia," *Journal of Great Lakes Research* 40 (2014): 226–46.

35. Litke, David W., "Review of Phosphorus Control Measures in the United States and Their Effects on Water Quality" (Denver, CO: US Department of the Interior, 1999).

36. Maccoux, Matthew J., Alice Dove, Sean M. Backus, and David M. Dolan, "Total and Soluble Reactive Phosphorus Loadings to Lake Erie: A Detailed Accounting by Year, Basin, Country, and Tributary," *Journal of Great Lakes Research* 42 (2016): 1151–65.

37. https://undark.org/2017/03/31/great-black-swamp-ohio-toledo/, accessed November 29, 2019.

38. Kaatz, Martin R., "The Black Swamp: A Study in Historical Geography," *Annals of the Association of American Geographers* 45 (1955): 1–35.

39. McCarthy, James F., "Lake Erie Dead Zone Threatens Cleveland Drinking Water," *Plain Dealer* (July 25, 2018), https://www.cleveland.com/metro/2018/07/lake_erie_dead_zone_poses_annu.html, accessed April 15, 2020.

40. Wood, Roslyn, "Acute Animal and Human Poisonings from Cyanotoxin Exposure—A Review of the Literature," *Environment International* 91 (2016): 276–82.

41. Bridgeman, Thomas B., Justin D. Chaffin, Douglas D. Kane, Joseph D. Conroy, Sarah E. Panek, and Patricia M. Armenio, "From River to Lake: Phosphorus Partitioning and Algal Community Compositional Changes in Western Lake Erie," *Journal of Great Lakes Research* 38 (2012): 90–97.

42. https://www.weather.gov/cle/LakeErieHAB, accessed December 2, 2019.

43. Steffen, Morgan M., Timothy W. Davis, R. Michael L. McKay, George S. Bullerjahn, Lauren E. Krausfeldt, Joshua M. A. Stough, Michelle L. Neitzey, et al., "Ecophysiological Examination of the Lake Erie *Microcystis* Bloom in 2014: Linkages between Biology and the Water Supply Shutdown of Toledo, OH," *Environmental Science & Technology* 51 (2017): 6745–55.

44. Newell, Silvia E., Timothy W. Davis, Thomas H. Johengen, Duane Gossiaux, Ashley Burtner, Danna Palladino, and Mark J. McCarthy, "Reduced Forms of Nitrogen Are a Driver of Non-Nitrogen-Fixing Harmful Cyanobacterial Blooms and Toxicity in Lake Erie," *Harmful Algae* 81 (2019): 86–93.

45. Gobler, Christopher J., JoAnn M. Burkholder, Timothy W. Davis, Matthew J. Harke, Tom Johengen, Craig A. Stow, and Dedmer B. Van de Waal, "The Dual Role of Nitrogen Supply in Controlling the Growth and Toxicity of Cyanobacterial Blooms," *Harmful Algae* 54 (2016): 87–97.

46. https://www.epa.gov/glwqa/us-action-plan-lake-erie, accessed December 2, 2019.

Chapter 8

1. Fullmer, June Z., *Young Humphry Davy: The Making of an Experimental Chemist* (Philadelphia: American Philosophical Society, 2000).

2. Vaquer-Sunyer, Raquel, and Carlos M. Duarte, "Thresholds of Hypoxia for Marine Biodiversity," *Proceedings of the National Academy of Sciences* 105 (2008): 15452–57.

3. Breitburg, Denise L., Darryl W. Hondorp, Lori A. Davias, and Robert J. Diaz, "Hypoxia, Nitrogen, and Fisheries: Integrating Effects across Local and Global Landscapes," *Annual Review of Marine Science* 1 (2009): 329–49.

4. Miller, Eric F., "Description of Conditions Preceding the 2011 Redondo Beach, California, Fish Kill," *Journal of Coastal Research* 30 (2014): 795–99.

5. Ouellet, Valérie, Marc Mingelbier, André Saint-Hilaire, and Jean Morin, "Frequency Analysis as a Tool for Assessing Adverse Conditions During a Massive Fish Kill in the St. Lawrence River, Canada," *Water Quality Research Journal* 45 (2010): 47–57.

6. Lowe, J. A., D. R. G. Farrow, A. S. Pait, S. J. Arenstam, and E. F. Lavan, "Fish Kills in Coastal Waters 1980–1989," (Washington, DC: National Oceanic and Atmospheric Administration, 1991).

7. Thronson, A., and A. Quigg, "Fifty-Five Years of Fish Kills in Coastal Texas," *Estuaries and Coasts* 31 (2008): 802–13.

8. May, Edwin B., "Extensive Oxygen Depletion in Mobile Bay, Alabama," *Limnology and Oceanography* 18 (1973): 353–66.

9. Bakun, Andrew, "Climate Change and Ocean Deoxygenation within Intensified Surface-Driven Upwelling Circulations," *Philosophical Transactions of the Royal Society A: Mathematical, Physical and Engineering Sciences* 375 (2017): 20160327.

10. Breitburg, Denise, Lisa A. Levin, Andreas Oschlies, Marilaure Grégoire, Francisco P. Chavez, Daniel J. Conley, Véronique Garçon, et al., "Declining Oxygen in the Global Ocean and Coastal Waters," *Science* 359 (2018): https://doi.org/10.1126/science.aam7240.

11. Wang, Junjie, Arthur H. W. Beusen, Xiaochen Liu, and Alexander F. Bouwman, "Aquaculture Production Is a Large, Spatially Concentrated Source of Nutrients in Chinese Freshwater and Coastal Seas," *Environmental Science & Technology* 54 (2020): 1464–74.

12. Rabalais, Nancy N., and R. Eugene Turner, "Gulf of Mexico Hypoxia: Past, Present, and Future," *Limnology and Oceanography Bulletin* 28 (2019): 117–24.

13. Chen, Yong, "Fish Resources of the Gulf of Mexico," in *Habitats and Biota of the Gulf of Mexico: Before the Deepwater Horizon Oil Spill, Volume 2: Fish Resources, Fisheries, Sea Turtles, Avian Resources, Marine Mammals, Diseases and Mortalities*, edited by C. Herb Ward (New York: Springer, 2017), pp. 869–1038.

14. The sum of both commercial and recreational fishing, according to https://www.fisheries.noaa.gov/, accessed August 23, 2019.

15. Ward, C. Herb, ed., *Habitats and Biota of the Gulf of Mexico: Before the Deepwater Horizon Oil Spill, Volume 2: Fish Resources, Fisheries, Sea Turtles, Avian Resources, Marine Mammals, Diseases and Mortalities* (New York: Springer, 2017).

16. https://www.audubon.org/conservation/gulf, accessed August 26, 2019. See also Burger, Joanna, "Avian Resources of the Northern Gulf of Mexico," in *Habitats and Biota of the Gulf of Mexico: Before the Deepwater Horizon Oil Spill: Volume 2: Fish Resources, Fisheries, Sea Turtles, Avian Resources, Marine Mammals, Diseases and Mortalities*, edited by C. Herb Ward (New York: Springer, 2017), pp. 1353–488.

17. Rose, Kenneth A., Aaron T. Adamack, Cheryl A. Murphy, Shaye E. Sable, Sarah E. Kolesar, J. Kevin Craig, Denise L. Breitburg, et al., "Does Hypoxia Have Population-Level Effects on Coastal Fish? Musings from the Virtual World," *Journal of Experimental Marine Biology and Ecology* 381 (2009): S188–S203.

18. https://gulfhypoxia.net/catching-fish-in-the-dead-zone/, accessed August 22, 2019.

19. Rabalais, N. N., R. E. Turner, and W. J. Wiseman, "Gulf of Mexico Hypoxia, aka 'the Dead Zone'," *Annual Review of Ecology and Systematics* 33 (2002): 235–63.

20. Rabalais, N. N., and M. M. Baustian, "Historical Shifts in Benthic Infaunal Diversity in the Northern Gulf of Mexico since the Appearance of Seasonally Severe Hypoxia," *Diversity* 12, 49 (2020): https://doi.org/10.3390/d12020049.

21. Breitburg et al., "Hypoxia, Nitrogen, and Fisheries."

22. Rabalais, Nancy. Interview with author, Baton Rouge, LA, January 30, 2019.

23. Scavia, Donald. Interview with author, Newark, DE, March 14, 2019.

24. Eero, Margit, Brian R. MacKenzie, Friedrich W. Köster, and Henrik Gislason, "Multi-Decadal Responses of a Cod (*Gadus morhua*) Population to Human-Induced Trophic Changes, Fishing, and Climate," *Ecological Applications* 21 (2011): 214–26.

25. Craig, J. K., "Aggregation on the Edge: Effects of Hypoxia Avoidance on the Spatial Distribution of Brown Shrimp and Demersal Fishes in the Northern Gulf of Mexico," *Marine Ecology Progress Series* 445 (2012): 75–95.

26. Rose, Kenneth A., Dimitri Gutiérrez, Denise Breitburg, Daniel Conley, J. Kevin Craig, Halley E. Froehlich, R. Jeyabaskaran, et al., "Impacts of Ocean Deoxygenation on Fisheries," in *Ocean Deoxygenation: Everyone's Problem: Causes, Impacts, Consequences and Solutions*, edited by D. Laffoley and J. M. Baxter (Gland, Switzerland: IUCN, 2019), pp. 519–44.

27. Langseth, Brian J., Kevin M. Purcell, Craig J. Kevin, Amy M. Schueller, Joseph W. Smith, Kyle W. Shertzer, Sean Creekmore, et al., "Effect of Changes in Dissolved Oxygen Concentrations on the Spatial Dynamics of the Gulf Menhaden Fishery in the Northern Gulf of Mexico," *Marine and Coastal Fisheries* 6 (2014): 223–34.

28. Purcell, Kevin M., J. Kevin Craig, James M. Nance, Martin D. Smith, and Lori S. Bennear, "Fleet Behavior Is Responsive to a Large-Scale Environmental Disturbance: Hypoxia Effects on the Spatial Dynamics of the Northern Gulf of Mexico Shrimp Fishery," *PLOS ONE* 12 (2017): e0183032.

29. Craig, "Aggregation on the Edge."

30. Zimmerman, Roger J., and James M. Nance, "Effects of Hypoxia on the Shrimp Fishery of Louisiana and Texas," in *Coastal Hypoxia: Consequences for Living Resources and Ecosystems*, edited by Nancy N. Rabalais and R. E. Turner (Washington, DC: American Geophysical Union, 2001), pp. 293–310.

31. https://gulfhypoxia.net/in-minn-gulf-shrimpers-meet-farmers-trying-to-save-their-catch/, accessed September 3, 2019.

32. Smith, Martin D., Atle Oglend, A. Justin Kirkpatrick, Frank Asche, Lori S. Bennear, J. Kevin Craig, and James M. Nance, "Seafood Prices Reveal Impacts of a Major Ecological Disturbance," *Proceedings of the National Academy of Sciences* 114 (2017): 1512–17.

33. Smith et al., "Seafood Prices."

34. Caffrey, Jane M., Martha Brown, W. Breck Tyler, and M Silberstein, eds., *Changes in a California Estuary: A Profile of Elkhorn Slough* (Moss Landing, CA: Elkhorn Slough Foundation, 2002).

35. https://www.seemonterey.com/things-to-do/parks/elkhorn-slough/ accessed September 5, 2019.

36. Hughes, Brent B., Matthew D. Levey, Monique C. Fountain, Aaron B. Carlisle, Francisco P. Chavez, and Mary G. Gleason, "Climate Mediates Hypoxic Stress on

Fish Diversity and Nursery Function at the Land–Sea Interface," *Proceedings of the National Academy of Sciences* 112 (2015): 8025–30.

37. Rose et al., "Impacts of Ocean Deoxygenation."
38. Rose, Kenneth. Telephone interview with author, September 18, 2019.
39. Roman, Michael R., Stephen B. Brandt, Edward D. Houde, and James J. Pierson, "Interactive Effects of Hypoxia and Temperature on Coastal Pelagic Zooplankton and Fish," *Frontiers in Marine Science* 6 (2019): https://doi.org/10.3389/fmars.2019.00139.
40. Breitburg et al., "Hypoxia, Nitrogen, and Fisheries."
41. Breitburg, Denise. Interview with author, Edgewater, MD, July 12, 2019.
42. Levitan, Richard, "How We Can Get Ahead of Covid-19," *New York Times*, April 26, 2020, p. 3, section SR.

Chapter 9

1. Henderson, Gabriel, and Roger Turner, "When Should Scientists Become Public Activists? The Oxygen Depletion Crisis," *Case Studies in the Environment* 2 (2018): 1–6.
2. Broecker, Wallace S., "Man's Oxygen Reserves," *Science* 168 (1970): 1537–38.
3. Keeling, Ralph F., and Stephen R. Shertz, "Seasonal and Interannual Variations in Atmospheric Oxygen and Implications for the Global Carbon Cycle," *Nature* 358 (1992): 723–27.
4. http://scrippso2.ucsd.edu/, accessed June 6, 2018.
5. Broecker, Wallace S., and Jeffrey P. Severinghaus, "Diminishing Oxygen," *Nature* 358 (1992): 710–11.
6. Freeland, Howard, "A Short History of Ocean Station Papa and Line P," *Progress in Oceanography* 75 (2007): 120–25.
7. Darwin, Charles, "The Beagle Letters," in *The Beagle Letters*, edited by Frederick Burkhardt and Conrad Martens (Cambridge: Cambridge University Press, 2008), p. 383.
8. Whitney, Frank A., Howard J. Freeland, and Marie Robert, "Persistently Declining Oxygen Levels in the Interior Waters of the Eastern Subarctic Pacific," *Progress in Oceanography* 75 (2007): 179–99.
9. World Ocean Database, National Centers for Environmental Information, https://www.nodc.noaa.gov.
10. Schmidtko, Sunke, Lothar Stramma, and Martin Visbeck, "Decline in Global Oceanic Oxygen Content During the Past Five Decades," *Nature* 542 (2017): 335–39.
11. Resplandy, L., "Will Ocean Zones with Low Oxygen Levels Expand or Shrink?," *Nature* 557 (2018): 314–15.
12. Stramma, Lothar, Gregory C. Johnson, Janet Sprintall, and Volker Mohrholz, "Expanding Oxygen-Minimum Zones in the Tropical Oceans," *Science* 320 (2008): 655–58.
13. Duteil, O., A. Oschlies, and C. W. Böning, "Pacific Decadal Oscillation and Recent Oxygen Decline in the Eastern Tropical Pacific Ocean," *Biogeosciences* 15 (2018): 7111–26.

14. Czeschel, Rena, Lothar Stramma, and Gregory C. Johnson, "Oxygen Decreases and Variability in the Eastern Equatorial Pacific," *Journal of Geophysical Research: Oceans* 117 (2012): https://doi.org/10.1029/2012jc008043.

15. Stramma, Lothar, and Sunke Schmidtko, "Global Evidence of Ocean Deoxygenation," in *Ocean Deoxygenation: Everyone's Problem: Causes, Impacts, Consequences and Solutions*, edited by D. Laffoley and J. M. Baxter (Gland, Switzerland: IUCN, 2019), pp. 25–36.

16. Dittmar, William, "Report on Researches into the Composition of Ocean-Water, Collected by H.M.S. Challenger, During the Years 1873–1876," in *Report of the Scientific Results of the Voyage of H.M.S. Challenger During the Years 1873–76: Physics and Chemistry Vol, 1*, edited by C. Wyville Thomson (London: Her Majesty's Stationery Office, 1884), pp. 1–265.

17. Schmidt, Johs, "On the Contents of Oxygen in the Ocean on Both Sides of Panama," *Science* 61 (1925): 592–93.

18. Ito, T., A. Nenes, M. S. Johnson, N. Meskhidze, and C. Deutsch, "Acceleration of Oxygen Decline in the Tropical Pacific over the Past Decades by Aerosol Pollutants," *Nature Geoscience* 9 (2016): 443–47.

19. Roemmich, Dean, John Church, John Gilson, Didier Monselesan, Philip Sutton, and Susan Wijffels, "Unabated Planetary Warming and Its Ocean Structure since 2006," *Nature Climate Change* 5 (2015): 240–45.

20. Ito, Takamitsu, Shoshiro Minobe, Matthew C. Long, and Curtis Deutsch, "Upper Ocean O_2 Trends: 1958–2015," *Geophysical Research Letters* 44 (2017): 4214–23.

21. Schmidtko et al., "Decline in Global Oceanic Oxygen."

22. Brewer, Peter G., and Edward T. Peltzer, "Ocean Chemistry, Ocean Warming, and Emerging Hypoxia: Commentary," *Journal of Geophysical Research: Oceans* 121 (2016): 3659–67.

23. Mazuecos, Ignacio P., Javier Arístegui, Evaristo Vázquez-Domínguez, Eva Ortega-Retuerta, Josep M. Gasol, and Isabel Reche, "Temperature Control of Microbial Respiration and Growth Efficiency in the Mesopelagic Zone of the South Atlantic and Indian Oceans," *Deep Sea Research Part I* 95 (2015): 131–38.

24. Levin, Lisa A., "Manifestation, Drivers, and Emergence of Open Ocean Deoxygenation," *Annual Review of Marine Science* 10 (2018): 229–60.

25. Powledge, Fred, "Chesapeake Bay Restoration: A Model of What?," *BioScience* 55 (2005): 1032–38.

26. Cheramie, Kristi Dykema, "The Scale of Nature: Modeling the Mississippi River," *Places Journal* (March 2011): https://doi.org/10.22269/110321.

27. Adler, Antony, *Neptune's Laboratory: Fantasy, Fear, and Science at Sea* (Cambridge, MA: Harvard University Press, 2019).

28. Stramma and Schmidtko, "Global Evidence of Ocean Deoxygenation."

29. Box, George E. P., and Norman Richard Draper, *Empirical Model-Building and Response Surfaces* (New York: Wiley, 1987), p. 424.

30. Grantham, Brian A., Francis Chan, Karina J. Nielsen, David S. Fox, John A. Barth, Adriana Huyer, Jane Lubchenco, et al., "Upwelling-Driven Nearshore Hypoxia Signals Ecosystem and Oceanographic Changes in the Northeast Pacific," *Nature* 429 (2004): 749–54.

31. Chan, F., J. A. Barth, J. Lubchenco, A. Kirincich, H. Weeks, W. T. Peterson, and B. A. Menge, "Emergence of Anoxia in the California Current Large Marine Ecosystem," *Science* 319 (2008): 920.

32. Email to author from Ervin Schumacker, Quinault Department of Fisheries, Quinault Indian Nation, United States, April 18, 2020.

33. Ren, Alice S., Fei Chai, Huijie Xue, David M. Anderson, and Francisco P. Chavez, "A Sixteen-Year Decline in Dissolved Oxygen in the Central California Current," *Scientific Reports* 8 (2018): 7290.

34. Chan, Francis, "Evidence for Ocean Deoxygenation and Its Patterns: Eastern Boundary Upwelling Systems," in *Ocean Deoxygenation: Everyone's Problem: Causes, Impacts, Consequences and Solutions*, edited by D. Laffoley and J. M. Baxter (Gland, Switzerland: IUCN, 2019), pp. 73–84.

35. Seibel, Brad A., "The Jumbo Squid, *Dosidicus Gigas* (Ommastrephidae), Living in Oxygen Minimum Zones II: Blood–Oxygen Binding," *Deep Sea Research Part II: Topical Studies in Oceanography* 95 (2013): 139–44.

36. Gilly, William F., J. Michael Beman, Steven Y. Litvin, and Bruce H. Robison, "Oceanographic and Biological Effects of Shoaling of the Oxygen Minimum Zone," *Annual Review of Marine Science* 5 (2013): 393–420.

37. Nigmatullin, Ch M., K. N. Nesis, and A. I. Arkhipkin, "A Review of the Biology of the Jumbo Squid *Dosidicus gigas* (Cephalopoda: Ommastrephidae)," *Fisheries Research* 54 (2001): 9–19.

38. Gilly et al., "Oceanographic and Biological Effects of Shoaling."

39. Koslow, J. A., R. Goericke, A. Lara-Lopez, and W. Watson, "Impact of Declining Intermediate-Water Oxygen on Deepwater Fishes in the California Current," *Marine Ecology Progress Series* 436 (2011): 207–18.

40. Worm, Boris, Marcel Sandow, Andreas Oschlies, Heike K. Lotze, and Ransom A. Myers, "Global Patterns of Predator Diversity in the Open Oceans," *Science* 309 (2005): 1365–69.

41. Breitburg, Denise L., Hannes Baumann, Inna M. Sokolova, and Christina A. Frieder, "Multiple Stressors—Forces That Combine to Worsen Deoxygenation and Its Effects," in *Ocean Deoxygenation: Everyone's Problem: Causes, Impacts, Consequences and Solutions*, edited by D. Laffoley and J. M. Baxter (Gland, Switzerland: IUCN, 2019), pp. 225–47.

42. Vaquer-Sunyer, Raquel, and Carlos M. Duarte, "Temperature Effects on Oxygen Thresholds for Hypoxia in Marine Benthic Organisms," *Global Change Biology* 17 (2011): 1788–97.

43. Deutsch, Curtis, Aaron Ferrel, Brad Seibel, Hans-Otto Pörtner, and Raymond B. Huey, "Climate Change Tightens a Metabolic Constraint on Marine Habitats," *Science* 348 (2015): 1132–35.

44. Davy, Humphry, *Researches, Chemical and Philosophical: Chiefly Concerning Nitrous Oxide, or Dephlogisticated Nitrous Air, and Its Respiration.* (London: J. Johnson, 1800), p. 489.

45. Davy, *Researches, Chemical and Philosophical*, p. 489.

46. Fowler, D., C. E. Steadman, D. Stevenson, M. Coyle, R. M. Rees, U. M. Skiba, M. A. Sutton, et al., "Effects of Global Change During the 21st Century on the Nitrogen Cycle," *Atmospheric Chemistry and Physics* 15 (2015): 13849–93.

Chapter 10

1. Levin, Kelly, Benjamin Cashore, Steven Bernstein, and Graeme Auld, "Overcoming the Tragedy of Super Wicked Problems: Constraining Our Future Selves to Ameliorate Global Climate Change," *Policy Sciences* 45 (2012): 123–52.
2. Khanna, Madhu, Benjamin M. Gramig, Evan H. DeLucia, Ximing Cai, and Praveen Kumar, "Harnessing Emerging Technologies to Reduce Gulf Hypoxia," *Nature Sustainability* 2 (2019): 889–91.
3. Streeter, M. T., and K. E. Schilling, "Effects of Golf Course Management on Subsurface Soil Properties in Iowa," *Soil* 4 (2018): 93–100.
4. Harris, L. A., C. L. S. Hodgkins, M. C. Day, D. Austin, J. M. Testa, W. Boynton, L. Van Der Tak, et al., "Optimizing Recovery of Eutrophic Estuaries: Impact of Destratification and Re-Aeration on Nutrient and Dissolved Oxygen Dynamics," *Ecological Engineering* 75 (2015): 470–83.
5. https://www.waternewsnetwork.com/new-oxygenation-system-improve-reservoir-water-quality/, accessed January 28, 2020.
6. Breitburg, Denise, Daniel J. Conley, Kirsten Isensee, Lisa A. Levin, Karin E. Limburg, and Phillip Williamson, "What Can We Do? Adaptation and Solutions to Declining Ocean Oxygen," in *Ocean Deoxygenation: Everyone's Problem: Causes, Impacts, Consequences and Solutions*, edited by D. Laffoley and J. M. Baxter (Gland, Switzerland: IUCN, 2019), pp. 545–62.
7. Clements, Laura, "The Strange Things That Happen to the Water of Cardiff Bay in a Heatwave," *WalesOnline*, July 19, 2018, https://www.walesonline.co.uk/news/wales-news/strange-things-happen-water-cardiff-14927326, accessed March 13, 2020.
8. Anomyous, "Oxygenation Barge Remediating Shanghai Waterway," *Journal—AWWA* 94 (2002): 62–63.
9. Data from Neil Dunlop, UK Environmental Agency, emailed to author, July 1, 2019.
10. Larsen, Sarah J., Kieryn L. Kilminster, Alessandra Mantovanelli, Zoë J. Goss, Georgina C. Evans, Lee D. Bryant, and Daniel F. McGinnis, "Artificially Oxygenating the Swan River Estuary Increases Dissolved Oxygen Concentrations in the Water and at the Sediment Interface," *Ecological Engineering* 128 (2019): 112–21.
11. Rydin, Emil, Linda Kumblad, Fredrik Wulff, and Per Larsson, "Remediation of a Eutrophic Bay in the Baltic Sea," *Environmental Science & Technology* 51 (2017): 4559–66.
12. Diaz, Robert J., Rutger Rosenberg, and Kersey Sturdivant, "Hypoxia in Estuaries and Semi-Enclosed Seas," in *Ocean Deoxygenation: Everyone's Problem: Causes, Impacts, Consequences and Solutions*, edited by D. Laffoley and J. M. Baxter (Gland, Switzerland: IUCN, 2019), pp. 85–102.

13. Hagy, J. D., W. R. Boynton, C. W. Keefe, and K. V. Wood, "Hypoxia in Chesapeake Bay, 1950-2001: Long-Term Change in Relation to Nutrient Loading and River Flow," *Estuaries* 27 (2004): 634–58.

14. http://www.iowadnr.gov/Environmental-Protection/Water-Quality/Water-Monitoring/Impaired-Waters, accessed March 13, 2020.

15. Jones, Chris, "It's Their Way or the Highway," September 16, 2019, https://www.iihr.uiowa.edu/cjones/its-their-way-or-the-highway, accessed May 5, 2020.

16. Townsend, Alan R., Robert W. Howarth, Fakhri A. Bazzaz, Mary S. Booth, Cory C. Cleveland, Sharon K. Collinge, Andrew P. Dobson, et al., "Human Health Effects of a Changing Global Nitrogen Cycle," *Frontiers in Ecology and the Environment* 1 (2003): 240–46.

17. Bryan, Nathan S., and Hans van Grinsven, "Chapter Three—the Role of Nitrate in Human Health," in *Advances in Agronomy*, edited by Donald L. Sparks (London: Academic Press, 2013), pp. 153–82.

18. Fowler, D., C. E. Steadman, D. Stevenson, M. Coyle, R. M. Rees, U. M. Skiba, M. A. Sutton, et al., "Effects of Global Change During the 21st Century on the Nitrogen Cycle," *Atmospheric Chemistry and Physics* 15 (2015): 13849–93.

19. Dodds, Walter K., Wes W. Bouska, Jeffrey L. Eitzmann, Tyler J. Pilger, Kristen L. Pitts, Alyssa J. Riley, Joshua T. Schloesser, et al., "Eutrophication of U.S. Freshwaters: Analysis of Potential Economic Damages," *Environmental Science & Technology* 43 (2009): 12–19.

20. Cullen, Art, *Storm Lake: A Chronicle of Change, Resilience, and Hope from a Heartland Newspaper* (New York: Viking, 2018).

21. Walton, Brett, "Nitrate Pollution Rising in Private Wells in Iowa," *Circle of Blue*, April 25, 2019, https://www.circleofblue.org/2019/world/nitrate-pollution-rising-in-private-wells-in-iowa, accessed March 13, 2020.

22. Schapiro, Mark, "In the Heart of the Corn Belt, an Uphill Battle for Clean Water," *YaleEnvironment360*, September 25, 2018, https://e360.yale.edu/features/in-the-heart-of-the-corn-belt-an-uphill-battle-for-clean-water-iowa, accessed March 13, 2020.

23. Elmer, MacKenzie, "Des Moines Water Works Won't Appeal Lawsuit," *Des Moines Register*, April 11, 2017, https://www.desmoinesregister.com/story/news/2017/04/11/des-moines-water-works-not-appeal-lawsuit/100321222, accessed March 13, 2020.

24. Norvell, Kim, "'We're All Lucky to Have Been Able to Know Him': Bill Stowe, Water Works CEO, Has Died," *Des Moines Register*, April 15, 2019, https://www.desmoinesregister.com/story/news/2019/04/14/des-moines-water-works-ceo-bill-stowe-pancreatic-cancer-obituary-iowa-nevada-grinnell-midamerican/3363533002/, accessed April 22, 2020.

25. Essman, Ellen, "The Des Moines Water Works Lawsuit: What's Happened, What's Next?," *Ohio Ag Manager*, February 22, 2017, https://u.osu.edu/ohioagmanager/2017/02/22/the-des-moines-water-works-lawsuit-whats-happened-whats-next/, accessed March 13, 2020.

26. Zulauf, C., and B. Brown, "Use of Tile, 2017 US Census of Agriculture," *Farmdoc Daily* 9 (2019): 141.

27. Valkama, Elena, Kirsi Usva, Merja Saarinen, and Jaana Uusi-Kämppä, "A Meta-Analysis on Nitrogen Retention by Buffer Zones," *Journal of Environmental Quality* 48 (2019): 270–79.

28. https://naturalresources.extension.iastate.edu/forestry/planning/buffer.html, accessed March 13, 2020.

29. Schilling, Keith E., Christopher S. Jones, Anthony Seeman, Eileen Bader, and Jennifer Filipiak, "Nitrate-Nitrogen Patterns in Engineered Catchments in the Upper Mississippi River Basin," *Ecological Engineering* 42 (2012): 1–9.

30. Scavia, Donald, Margaret Kalcic, Rebecca Logsdon Muenich, Jennifer Read, Noel Aloysius, Isabella Bertani, Chelsie Boles, et al., "Multiple Models Guide Strategies for Agricultural Nutrient Reductions," *Frontiers in Ecology and the Environment* 15 (2017): 126–32.

31. Hunt, Natalie D., Jason D. Hill, and Matt Liebman, "Cropping System Diversity Effects on Nutrient Discharge, Soil Erosion, and Agronomic Performance," *Environmental Science & Technology* 53 (2019): 1344–52.

32. Sharpley, Andrew, Matthew J. Helmers, Peter Kleinman, Kevin King, April Leytem, and Nathan Nelson, "Managing Crop Nutrients to Achieve Water Quality Goals," *Journal of Soil and Water Conservation* 74 (2019): 91A–101A.

33. Jones, Chris, "Make America MRTN Again," June 21, 2019, https://www.iihr.uiowa.edu/cjones/make-america-mrtn-again, accessed March 13, 2020.

34. Zhang, Xin, Eric A. Davidson, Denise L. Mauzerall, Timothy D. Searchinger, Patrice Dumas, and Ye Shen, "Managing Nitrogen for Sustainable Development," *Nature* 528 (2015): 51–59.

35. https://www.fsa.usda.gov/programs-and-services/conservation-programs/conservation-reserve-program/index, accessed March 13, 2020.

36. https://www.fsa.usda.gov/programs-and-services/conservation-programs/reports-and-statistics/conservation-reserve-program-statistics/index, accessed September 24, 2020.

37. Mitsch, William J., "Solving Lake Erie's Harmful Algal Blooms by Restoring the Great Black Swamp in Ohio," *Ecological Engineering* 108 (2017): 406–13.

38. Jones, Chris, "Pipe Dreams," June 14, 2019, https://www.iihr.uiowa.edu/cjones/pipe-dreams, accessed March 13, 2020.

39. Keiser, David A., Catherine L. Kling, and Joseph S. Shapiro, "The Low but Uncertain Measured Benefits of US Water Quality Policy," *Proceedings of the National Academy of Sciences* 116 (2019): 5262–69.

40. Tallis, Heather, Stephen Polasky, Jessica Hellmann, Nathaniel P. Springer, Rich Biske, Dave DeGeus, Randal Dell, et al., "Five Financial Incentives to Revive the Gulf of Mexico Dead Zone and Mississippi Basin Soils," *Journal of Environmental Management* 233 (2019): 30–38.

41. Scavia, D., "Nutrient Pollution: Voluntary Steps Are Failing to Shrink Algae Blooms and Dead Zones," *The Conversation*, July 31, 2017, https://theconversation.com/nutrient-pollution-voluntary-steps-are-failing-to-shrink-algae-blooms-and-dead-zones-81249, accessed March 13, 2020.

42. Environmental Working Group Conservation database (https://conservation.ewg.org/).

43. Williamson, James M., "The Role of Information and Prices in the Nitrogen Fertilizer Management Decision: New Evidence from the Agricultural Resource Management Survey," *Journal of Agricultural and Resource Economics* 36 (2011): 552–72.

44. https://dnr.maryland.gov/criticalarea/Pages/default.aspx, accessed September 24, 2020.

45. https://www.deq.virginia.gov/Programs/Water/ChesapeakeBay/ ChesapeakeBayPreservationAct.aspx, accessed January 2, 2020.

46. Liu, Jian, Peter J. A. Kleinman, Helena Aronsson, Don Flaten, Richard W. McDowell, Marianne Bechmann, Douglas B. Beegle, et al., "A Review of Regulations and Guidelines Related to Winter Manure Application," *Ambio* 47 (2018): 657–70.

47. Wang, Mengru, Lin Ma, Maryna Strokal, Wenqi Ma, Xuejun Liu, and Carolien Kroeze, "Hotspots for Nitrogen and Phosphorus Losses from Food Production in China: A County-Scale Analysis," *Environmental Science & Technology* 52 (2018): 5782–91.

48. Crawford, John T., Edward G. Stets, and Lori A. Sprague, "Network Controls on Mean and Variance of Nitrate Loads from the Mississippi River to the Gulf of Mexico," *Journal of Environmental Quality* 48 (2019): 1789–99.

49. Boesch, Donald F., "Barriers and Bridges in Abating Coastal Eutrophication," *Frontiers in Marine Science* 6 (2019): https://doi.org/10.3389/fmars.2019.00123.

50. Scavia, "Nutrient Pollution."

51. Kronvang, Brian, Hans E. Andersen, Christen Børgesen, Tommy Dalgaard, Søren E. Larsen, Jens Bøgestrand, and Gitte Blicher-Mathiasen, "Effects of Policy Measures Implemented in Denmark on Nitrogen Pollution of the Aquatic Environment," *Environmental Science & Policy* 11 (2008): 144–52.

52. Carstensen, Jacob, email to author, February 26, 2020.

53. Riemann, Bo, Jacob Carstensen, Karsten Dahl, Henrik Fossing, Jens W. Hansen, Hans H. Jakobsen, Alf B. Josefson, et al., "Recovery of Danish Coastal Ecosystems after Reductions in Nutrient Loading: A Holistic Ecosystem Approach," *Estuaries and Coasts* 39 (2016): 82–97.

54. Laurent, Arnaud, Katja Fennel, Dong S. Ko, and John Lehrter, "Climate Change Projected to Exacerbate Impacts of Coastal Eutrophication in the Northern Gulf of Mexico," *Journal of Geophysical Research: Oceans* 123 (2018): 3408–26.

55. Apuzzo, Matt, Selam Gebrekidan, Agustin Armendariz, and Jin Wu, "Killer Slime, Dead Birds, an Expunged Map: The Dirty Secrets of European Farm Subsidies," *New York Times*, Dec. 25, 2019, https://www.nytimes.com/interactive/2019/12/25/world/europe/farms-environment.html?action=click&module=Top%20 Stories&pgtype=Homepage, accessed March 13, 2020.

56. HELCOM, "Sources and Pathways of Nutrients to the Baltic Sea," (Helsinki, Finland, 2018) https://helcom.fi/media/publications/BSEP153.pdf.

57. Murray, Ciarán J., Bärbel Müller-Karulis, Jacob Carstensen, Daniel J. Conley, Bo G. Gustafsson, and Jesper H. Andersen, "Past, Present and Future Eutrophication Status of the Baltic Sea," *Frontiers in Marine Science* 6 (2019) 10.3389/fmars.2019.00002:

58. Hoekman, S. Kent, Amber Broch, and Xiaowei Liu, "Environmental Implications of Higher Ethanol Production and Use in the U.S.: A Literature Review. Part I—Impacts on Water, Soil, and Air Quality," *Renewable and Sustainable Energy Reviews* 81 (2018): 3140–58.

59. Chen, Xiaoguang, and Madhu Khanna, "Effect of Corn Ethanol Production on Conservation Reserve Program Acres in the US," *Applied Energy* 225 (2018): 124–34.

60. Secchi, Silvia, Philip W. Gassman, Manoj Jha, Lyubov Kurkalova, and Catherine L. Kling, "Potential Water Quality Changes Due to Corn Expansion in the Upper Mississippi River Basin," *Ecological Applications* 21 (2011): 1068–84.

61. Boesch, "Barriers and Bridges."

62. Poore, J., and T. Nemecek, "Reducing Food's Environmental Impacts through Producers and Consumers," *Science* 360 (2018): 987–92.

63. Toti, Elisabetta, Carla Di Mattia, and Mauro Serafini, "Metabolic Food Waste and Ecological Impact of Obesity in FAO World's Region," *Frontiers in Nutrition* 6 (2019): https://doi.org/10.3389/fnut.2019.00126.

64. https://www.who.int/gho/ncd/risk_factors/overweight_obesity/obesity_adults/en/, accessed May 8, 2020.

65. Clark, Michael A., Marco Springmann, Jason Hill, and David Tilman, "Multiple Health and Environmental Impacts of Foods," *Proceedings of the National Academy of Sciences* 116 (2019): 23357–62.

66. US Department of Agriculture Economic Research Service, https://www.ers.usda.gov/data-products/sugar-and-sweeteners-yearbook-tables/, accessed September 24, 2020.

67. Muth, Mary K., Catherine Birney, Amanda Cuéllar, Steven M. Finn, Mark Freeman, James N. Galloway, Isabella Gee, et al., "A Systems Approach to Assessing Environmental and Economic Effects of Food Loss and Waste Interventions in the United States," *Science of The Total Environment* 685 (2019): 1240–54.

68. Scherhaufer, Silvia, Graham Moates, Hanna Hartikainen, Keith Waldron, and Gudrun Obersteiner, "Environmental Impacts of Food Waste in Europe," *Waste Management* 77 (2018): 98–113.

Selected Bibliography

General reviews and books

Breitburg, Denise, Lisa A. Levin, Andreas Oschlies, Marilaure Grégoire, Francisco P. Chavez, Daniel J. Conley, Véronique Garçon, et al. "Declining Oxygen in the Global Ocean and Coastal Waters." *Science* 359 (2018): https://doi.org/10.1126/science. aam7240.

Fennel, Katja, and Jeremy M. Testa. "Biogeochemical Controls on Coastal Hypoxia." *Annual Review of Marine Science* 11 (2019): 105–30.

Laffoley, Daniel D'A, and J. M. Baxter, eds. *Ocean Deoxygenation: Everyone's Problem— Causes, Impacts, Consequences and Solutions* (Gland, Switzerland: IUCN, 2019).

Vaquer-Sunyer, Raquel, and Carlos M. Duarte. "Thresholds of Hypoxia for Marine Biodiversity." *Proceedings of the National Academy of Sciences* 105 (2008): 15452–57.

Websites

Baltic Marine Environment Protection Commission (HELCOM): https://helcom.fi

Chesapeake Bay Foundation: https://www.cbf.org/

Global Ocean Oxygen Network (GO$_2$ NE): https://en.unesco.org/go2ne

Gulf of Mexico hypoxia: https://gulfhypoxia.net

Ocean deoxygenation: https://www.iucn.org/theme/marine-and-polar/our-work/climate-change-and-oceans/ocean-deoxygenation

Chapter 1 The Great Stinks

Halliday, Stephen. *The Great Stink of London: Sir Joseph Bazalgette and the Cleansing of the Victorian Metropolis* (Stroud, England: Sutton, 1999).

Kauffman, Gerald J., Andrew R. Homsey, Andrew C. Belden, and Jessica Rittler Sanchez. "Water Quality Trends in the Delaware River Basin (USA) from 1980 to 2005." *Environmental Monitoring and Assessment* 177 (2011): 193–225.

Keiser, David A., and Joseph S. Shapiro. "Consequences of the Clean Water Act and the Demand for Water Quality." *The Quarterly Journal of Economics* 134 (2019): 349–96.

Sedlak, David L. *Water 4.0: The Past, Present, and Future of the World's Most Vital Resource* (New Haven: Yale University Press, 2014).

Wheeler, Alwyne C. *The Tidal Thames: The History of a River and Its Fishes* (London: Routledge & Kegan Paul, 1979).

Chapter 2 Dead Zones Discovered in Coastal Waters

Altieri, Andrew H., Seamus B. Harrison, Janina Seemann, Rachel Collin, Robert J. Diaz, and Nancy Knowlton. "Tropical Dead Zones and Mass Mortalities on Coral Reefs." *Proceedings of the National Academy of Sciences* 114 (2017): 3660–65.

Chen, Chung-Chi, Gwo-Ching Gong, and Fuh-Kwo Shiah. "Hypoxia in the East China Sea: One of the Largest Coastal Low-Oxygen Areas in the World." *Marine Environmental Research* 64 (2007): 399–408.

Fonselius, Stig, and Jorge Valderrama. "One Hundred Years of Hydrographic Measurements in the Baltic Sea." *Journal of Sea Research* 49 (2003): 229–41.

Turner, R. E., and R. L. Allen. "Bottom Water Oxygen Concentration in the Mississippi River Delta Bight." *Contributions in Marine Science* 25 (1982): 161–72.

Chapter 3 Coastal Dead Zones in the Past

Carstensen, Jacob, Jesper H. Andersen, Bo G. Gustafsson, and Daniel J. Conley. "Deoxygenation of the Baltic Sea During the Last Century." *Proceedings of the National Academy of Sciences* 111 (2014): 5628–33.

Kralj, M., M. Lipizer, B. Čermelj, M. Celio, C. Fabbro, F. Brunetti, J. Francé, P. Mozetič, and M. Giani. "Hypoxia and Dissolved Oxygen Trends in the Northeastern Adriatic Sea (Gulf of Trieste)." *Deep Sea Research Part II: Topical Studies in Oceanography* 164 (2019): 74–88.

Matli, V. Rohith Reddy, Shiqi Fang, Joseph Guinness, Nancy N. Rabalais, J. Kevin Craig, and Daniel R. Obenour. "Space-Time Geostatistical Assessment of Hypoxia in the Northern Gulf of Mexico." *Environmental Science & Technology* 52 (2018): 12484–93.

Osterman, L. E., R. Z. Poore, and P. W. Swarzenski. "The Last 1000 Years of Natural and Anthropogenic Low-Oxygen Bottom-Water on the Louisiana Shelf, Gulf of Mexico." *Marine Micropaleontology* 66 (2008): 291–303.

Chapter 4 What Happened in 1950?

Capet, A., E. V. Stanev, J. M. Beckers, J. W. Murray, and M. Grégoire. "Decline of the Black Sea Oxygen Inventory." *Biogeosciences* 13 (2016): 1287–97.

Harding, L. W., C. L. Gallegos, E. S. Perry, W. D. Miller, J. E. Adolf, M. E. Mallonee, and H. W. Paerl. "Long-Term Trends of Nutrients and Phytoplankton in Chesapeake Bay." *Estuaries and Coasts* 39 (2016): 664–81.

Norbäck Ivarsson, Lena, Thomas Andrén, Matthias Moros, Thorbjørn Joest Andersen, Mikael Lönn, and Elinor Andrén. "Baltic Sea Coastal Eutrophication in a Thousand Year Perspective." *Frontiers in Environmental Science* 7 (2019): https://doi.org/10.3389/fenvs.2019.00088.

Rabalais, Nancy N., and R. Eugene Turner. "Gulf of Mexico Hypoxia: Past, Present, and Future." *Limnology and Oceanography Bulletin* 28 (2019): 117–24.

Zillén, Lovisa, Daniel J. Conley, Thomas Andrén, Elinor Andrén, and Svante Björck. "Past Occurrences of Hypoxia in the Baltic Sea and the Role of Climate Variability, Environmental Change and Human Impact." *Earth-Science Reviews* 91 (2008): 77–92.

Chapter 5 Giving the Land a Kick

Cao, P., C. Lu, and Z. Yu. "Historical Nitrogen Fertilizer Use in Agricultural Ecosystems of the Contiguous United States During 1850–2015: Application Rate, Timing, and Fertilizer Types." *Earth System Science Data* 10 (2018): 969–84.

Robertson, Dale M., and David A. Saad. "Sparrow Models Used to Understand Nutrient Sources in the Mississippi/Atchafalaya River Basin." *Journal of Environmental Quality* 42 (2013): 1422–40.

Turner, R. Eugene, and Nancy N. Rabalais. "Linking Landscape and Water Quality in the Mississippi River Basin for 200 Years." *BioScience* 53 (2003): 563–72.

Yu, ChaoQing, Xiao Huang, Han Chen, H. Charles J. Godfray, Jonathon S. Wright, Jim W. Hall, Peng Gong, et al. "Managing Nitrogen to Restore Water Quality in China." *Nature* 567 (2019): 516–20.

Chapter 6 Liebig's Law and Haber's Tragedy

Brock, W. H. *Justus Von Liebig: The Chemical Gatekeeper* (Cambridge: Cambridge University Press, 1997).

Schindler, David W. "The Dilemma of Controlling Cultural Eutrophication of Lakes." *Proceedings of the Royal Society B: Biological Sciences* 279 (2012): 4322–33.

Smil, Vaclav. *Enriching the Earth: Fritz Haber, Carl Bosch, and the Transformation of World Food Production* (Cambridge, MA: MIT Press, 2001).

Turner, R. E., and N. N. Rabalais. "Nitrogen and Phosphorus Phytoplankton Growth Limitation in the Northern Gulf of Mexico." *Aquatic Microbial Ecology* 68 (2013): 159–69.

Chapter 7 The Case for Phosphorus

Fennel, K., and A. Laurent. "N and P as Ultimate and Proximate Limiting Nutrients in the Northern Gulf of Mexico: Implications for Hypoxia Reduction Strategies." *Biogeosciences* 15 (2018): 3121–31.

Jenny, Jean-Philippe, Pierre Francus, Alexandre Normandeau, François Lapointe, Marie-Elodie Perga, Antti Ojala, Arndt Schimmelmann, and Bernd Zolitschka. "Global Spread of Hypoxia in Freshwater Ecosystems During the Last Three Centuries Is Caused by Rising Local Human Pressure." *Global Change Biology* 22 (2016): 1481–89.

Redfield, Alfred C. "The Biological Control of Chemical Factors in the Environment." *American Scientist* 46 (1958): 205–21.

Steffen, Morgan M., Timothy W. Davis, R. Michael L. McKay, George S. Bullerjahn, Lauren E. Krausfeldt, Joshua M. A. Stough, Michelle L. Neitzey, et al. "Ecophysiological Examination of the Lake Erie *Microcystis* Bloom in 2014: Linkages between Biology and the Water Supply Shutdown of Toledo, OH." *Environmental Science & Technology* 51 (2017): 6745–55.

Chapter 8   Fish and Fisheries

Breitburg, Denise L., Darryl W. Hondorp, Lori A. Davias, and Robert J. Diaz. "Hypoxia, Nitrogen, and Fisheries: Integrating Effects across Local and Global Landscapes." *Annual Review of Marine Science* 1 (2009): 329–49.

Craig, J. K. "Aggregation on the Edge: Effects of Hypoxia Avoidance on the Spatial Distribution of Brown Shrimp and Demersal Fishes in the Northern Gulf of Mexico." *Marine Ecology Progress Series* 445 (2012): 75–95.

Hughes, Brent B., Matthew D. Levey, Monique C. Fountain, Aaron B. Carlisle, Francisco P. Chavez, and Mary G. Gleason. "Climate Mediates Hypoxic Stress on Fish Diversity and Nursery Function at the Land–Sea Interface." *Proceedings of the National Academy of Sciences* 112 (2015): 8025–30.

Smith, Martin D., Atle Oglend, A. Justin Kirkpatrick, Frank Asche, Lori S. Bennear, J. Kevin Craig, and James M. Nance. "Seafood Prices Reveal Impacts of a Major Ecological Disturbance." *Proceedings of the National Academy of Sciences* 114 (2017):   1512–17.

Chapter 9   Dead Zones in the Oceans

Levin, Lisa A. "Manifestation, Drivers, and Emergence of Open Ocean Deoxygenation." *Annual Review of Marine Science* 10 (2018): 229–60.

Ren, Alice S., Fei Chai, Huijie Xue, David M. Anderson, and Francisco P. Chavez. "A Sixteen-Year Decline in Dissolved Oxygen in the Central California Current." *Scientific Reports* 8 (2018): 7290.

Resplandy, L. "Will Ocean Zones with Low Oxygen Levels Expand or Shrink?" *Nature* 557 (2018): 314–15.

Schmidtko, Sunke, Lothar Stramma, and Martin Visbeck. "Decline in Global Oceanic Oxygen Content During the Past Five Decades." *Nature* 542 (2017): 335–39.

Vaquer-Sunyer, Raquel, and Carlos M. Duarte. "Temperature Effects on Oxygen Thresholds for Hypoxia in Marine Benthic Organisms." *Global Change Biology* 17 (2011): 1788–97.

Chapter 10   Reviving Dead Zones

Boesch, Donald F. "Barriers and Bridges in Abating Coastal Eutrophication." *Frontiers in Marine Science* 6 (2019): https://doi.org/10.3389/fmars.2019.00123.

Clark, Michael A., Marco Springmann, Jason Hill, and David Tilman. "Multiple Health and Environmental Impacts of Foods." *Proceedings of the National Academy of Sciences* 116 (2019): 23357–62.

Laurent, Arnaud, Katja Fennel, Dong S. Ko, and John Lehrter. "Climate Change Projected to Exacerbate Impacts of Coastal Eutrophication in the Northern Gulf of Mexico." *Journal of Geophysical Research: Oceans* 123 (2018): 3408–26.

Scavia, Donald, Margaret Kalcic, Rebecca Logsdon Muenich, Jennifer Read, Noel Aloysius, Isabella Bertani, Chelsie Boles, et al. "Multiple Models Guide Strategies for Agricultural Nutrient Reductions." *Frontiers in Ecology and the Environment* 15 (2017): 126–32.

Index

For the benefit of digital users, indexed terms that span two pages (e.g., 52–53) may, on occasion, appear on only one of those pages.